Catalysis in Chemistry

Methuen Studies in Science

GENERAL EDITOR   J. M. Gregory M.A., D.Phil., Winchester College

CONSULTANT EDITORS   B. E. Dawson B.Sc., Ph.D., King's College, London

R. Gliddon B.Sc., Ph.D., Clifton College, Bristol

This series provides students with concise, introductory surveys of important topics in the physical, chemical and biological sciences. The series is designed to assist students preparing for entry to university or college, and to meet the needs of university students preparing for more advanced studies.

Methuen Studies in Science

Catalysis in Chemistry

A. J. B. ROBERTSON

Professor of Chemistry,
King's College, University of London

47618

Methuen Educational Ltd

LONDON · TORONTO · SYDNEY · WELLINGTON

First published 1972
by Methuen Educational Ltd
11 New Fetter Lane, London EC4
© 1972 by A. J. B. Robertson
Printed in Great Britain by
William Clowes & Sons Ltd
London, Colchester and Beccles

SBN 423 84370 2   non net
SBN 423 84380 X   net

Distributed in the U.S.A. by
HARPER & ROW PUBLISHERS, INC.
BARNES & NOBLE IMPORT DIVISION

Contents

Preface

Applications of catalysis in chemistry extend over a vast field, and detailed theories and discussions of catalytic phenomena abound in the chemical literature. My brief account attempts to give only some general idea of the field, like an impressionistic sketch, and is undoubtedly influenced by personal and subjective interests.

I am, as with all my scientific writing over the past years, under a deep debt of gratitude to my wife for her unfailing help in so many ways.

A. J. B. Robertson

King's College
London

1 Historical introduction

Catalysis and catalysts

A catalyst is a substance which alters the velocity of a chemical reaction, although at
the end of the reaction the catalyst is still present and has not undergone any perma-
nent chemical change. The physical state of the catalyst may, however, change.
Examples of such a change are provided by manganese dioxide, which when used as
a catalyst for the reaction of the decomposition by heat of potassium chlorate is
usually in a finer state of subdivision at the end of the reaction, and by a smooth
platinum wire, the surface of which is roughened when the wire is used as a catalyst
for the oxidation of ammonia. A catalyst may be gaseous, liquid or solid. It is cus-
tomary to confine the term 'catalyst' to a substance which speeds up a chemical
reaction, although for many reactions substances are known which reduce the velocity of reaction. Such substances are usually called negative catalysts. It is usually
possible to find, for any particular reaction, a catalyst which will speed up the reaction, but to find a negative catalyst for a particular reaction may be difficult and
perhaps not even possible.

The general phenomenon of the action of catalysts is known as catalysis. The
subject of catalysis is of interest and importance in many ways. Thus, the phenomena present many mysteries which are not yet understood and so provide a challenge
to the chemist; reactions in living organisms involve complicated processes of catalysis
by enzymes, the study of which may throw light on problems of health, disease, and
ageing; and most of the reactions which the chemical industries use involve catalytic
actions.

Catalysis may be homogeneous or heterogeneous. In homogeneous catalysis the
catalyst and the reactant or reactants are in the same phase which in general will be
a liquid or gaseous phase. An example is the decomposition of liquid hydrogen peroxide or its solution accelerated by a catalyst soluble in the liquid. In heterogeneous
catalysis the catalyst and the reactant or reactants are in different phases — very
often the catalyst is a solid and the reactants are in a liquid or a gaseous phase. The
actual catalytic reaction then takes place at the surface of the solid where it makes
contact with the liquid or gas. We may say that the reaction occurs at an interface,
or at a phase boundary. The demarcation between homogeneous and heterogeneous
catalysis is, however, not always clear. Thus many reactions of small molecules in
solution are catalysed by enzymes consisting of very large molecules, and the small
molecule actually reacts at the surface of the enzyme which is exposed to the solution, so that the actual reaction is in effect at an interface and we could regard it

as an example of heterogeneous catalysis. Normally, however, we would regard the solution itself as being homogeneous, that is having uniform properties throughout. So the catalysis could also be described as homogeneous. Again, many reactions are catalysed by colloidal suspensions of metals, and there is no clearly defined division between suspensions and solutions. Indeed, the whole idea of a homogeneous phase is only a clear concept when we take a macroscopic view of matter — when we look at a solution in terms of individual molecules, or take a molecular view, it is clear that properties must vary from place to place and with time. The density of a liquid, for example, in a region of space of fixed size containing only a few molecules must fluctuate with time as molecules move in or out of the region. In fact, concepts and definitions introduced for convenience of the human mind do not always fully correspond to the complexities of natural phenomena.

Early work with catalysts

The basic concepts of catalysis mentioned in the preceding section were not formulated until the nineteenth century, although catalytic processes have been used by mankind for some thousands of years. The process of fermentation, known since primitive times, is involved in the conversion of fruit juice into wine, and of malt into beer, and in the making of vinegar, bread and cheese. The complex chemical reactions which occur during fermentation depend upon catalytically active enzymes produced by micro-organisms.

Metallic catalysts were used in laboratory operations before 1800 by J. Priestley and by D. van Marum, both of whom investigated the dehydrogenation of alcohol in contact with various heated metals. However, at that time the concept of catalysis as a natural phenomenon had not arisen and it seems likely that these investigators regarded the metals merely as a source of heat. In 1813, L. G. Thenard discovered that ammonia is decomposed into nitrogen and hydrogen when passed over various red-hot metals, and ten years later Thenard with P. L. Dulong found that the metals iron, copper, silver, gold and platinum are effective in decomposing ammonia. For given conditions, the ability to bring about the decomposition varied from one metal to another, decreasing in the order given above. We would now describe this variation of catalytic ability as a 'pattern of catalytic activity'. There is much debate at the present time amongst physical chemists interested in catalysis concerning the fundamental reasons for the existence of a pattern of catalytic activity when the efficiency is found for a number of different solid catalysts accelerating a given reaction.

Further important early studies of catalysts, this time in a liquid system, were made by Thenard after his discovery of hydrogen peroxide, described in a scientific paper published in 1818. He prepared hydrogen peroxide from barium peroxide by the action on it of nitric or hydrochloric acid. The interest of Thenard in barium peroxide derived from the early work of H. Davy 'On some Chemical Agencies of Electricity' which was the title of Davy's first Bakerian Lecture to the Royal Society in 1806. In that year the 3,000 francs prize of Napoleon I was awarded to Davy by the French awarding Committee, despite the existence of a state of war between England and France, for the work described in this lecture, subsequently published as a scientific paper. Davy at the Royal Institution had the use of a large battery (or voltaic pile, as it was then called) and Napoleon, noting the vital part this had played in the Davy research, and apparently somewhat peeved that his prize had been awarded to an Englishman, ordered two large voltaic piles to be built. Meanwhile Davy, in 1807 announced the discovery of sodium and potassium by electrolysis of the moist hydroxides, and this led Thenard to investigate the alkaline earth oxides, probably with the initial aim of electrolysing them. It was this work which led to his investigation of barium peroxide and the discovery of hydrogen peroxide.

Thenard observed that hydrogen peroxide decomposed when in contact with solids, such as charcoal, various metals, and various metal oxides. Some oxides were reduced, e.g. silver oxide to silver and lead dioxide (lead(IV) oxide) to lead monoxide (lead(II) oxide), but others were not chemically changed although they caused the peroxide to decompose. The action of metals in bringing about the decomposition was found to be more vigorous as the metal became in a finer state of subdivision.

A small concentration of a suitable catalyst dissolved in aqueous hydrogen peroxide can increase the decomposition rate of the peroxide by a very large factor. This factor may, for example, be a million or even a thousand million, depending on the chemical nature of the catalyst and on other important conditions, such as temperature and the concentration of the solution. Such huge increases in reaction rate brought about by a catalyst are quite usual.

Early work on heterogeneous catalytic oxidation

H. Davy after returning from his European tour in 1815 was asked to investigate the explosions in coal mines which were initiated by the lights used in mines. These researches led to the development of the Davy miner's safety-lamp, in which a luminous flame was surrounded by a cylinder of wire gauze which prevented the propagation of flame from inside the cylinder to a potentially explosive mixture outside. During these researches, in which Davy was assisted by M. Faraday, a logical step was to investigate the phenomena arising when a combustible gas mixture is in contact with a metal wire. This was done, and Davy discovered in 1817 that mixtures of combustible vapours and air could be oxidised with platinum and palladium wires without production of flame, but with the generation of enough heat to keep the wire incandescent. In fact, a hot wire introduced into the combustible gas mixture immediately became incandescent and remained so for a long time. Copper, silver, iron, gold and zinc were, however, ineffective in contrast to platinum and palladium. This is one of the earliest recorded patterns of catalytic activity.

M. Faraday in the Royal Institution laboratory later continued with the investigation of heterogeneous catalysis, and in 1834 he published a very important paper on the oxidation of hydrogen to water on a platinum surface. During his work on electrolysis, which led to the discovery of the laws of electrolysis now called Faraday's Laws, Faraday observed a spontaneous combination of hydrogen and oxygen on platinum electrodes at room temperature, and he concluded that this reaction, and similar reactions between nitrous oxide (dinitrogen oxide) and hydrogen and nitric oxide (nitrogen oxide) and hydrogen, always arose if the platinum surface was clean. Faraday in his paper reviewed previous work and proposed the theory, still held, that adsorption of gases on solid surfaces is intimately connected with any catalytic action of these surfaces on the gases. By adsorption of molecules is meant an adhesion of gas molecules to the solid surface so that a layer of molecules, more or less firmly held, exists at the surface.

The possibility of industrial application of catalytic oxidation was appreciated by that time, and in 1831 P. Phillips took out a patent for the production of sulphuric acid by the oxidation of sulphur dioxide on a platinum catalyst. However, this process could not compete at that time with the lead chamber process.

The recognition of catalysis as a natural phenomenon

At the time of the researches on catalysis by Faraday, the noted Swedish chemist J. J. Berzelius prepared for the Swedish Academy of Sciences an Annual Report reviewing the progress of chemistry, and in his report published in 1836 Berzelius coordinated a number of earlier observations, including those of Thenard and Davy, on both homogeneous and heterogeneous catalytic actions. He concluded: 'It is then proved that several simple and compound bodies soluble and insoluble, have the property of exercising on other bodies an action very different from chemical affinity. By means of this action they produce, in these bodies, decompositions of their elements and different recombinations of these same elements to which they themselves remain indifferent.' Berzelius proposed the introduction of the term 'catalysis' to describe the action, which he ascribed to a 'catalytic force' common to organic and inorganic nature. This report by Berzelius is probably the first recognition of catalytic action as a natural phenomenon of wide occurrence.

Catalysis and the equilibrium position in a reversible reaction

W. Ostwald emphasised that a catalyst cannot change the equilibrium position in a reversible reaction. Consider, for example, a reversible reaction $A \rightleftharpoons B$ which proceeds from left to right with evolution of heat. Suppose a catalyst existed which altered the equilibrium, causing it to move, on introduction of the catalyst, from left to right with production of more B. Then on introducing the catalyst, heat would be evolved, and this could be extracted from the system. The catalyst could then be removed, when reaction would go from right to left with absorption of heat. The system could be allowed to absorb this heat from the surroundings. Let us assume that the catalyst can be put in and out of the system without any work being done. Then by alternately putting the catalyst in and out of the system heat could be extracted from the surroundings by the reversible reaction without any work being done. But this would be a direct contradiction of the second law of thermodynamics. We therefore deduce that a catalyst cannot alter the equilibrium position in a reversible reaction. Again, one could consider a reversible reaction occurring with a change of volume in an apparatus fitted with a hypothetical frictionless and weightless piston, when the alternate introduction and removal of a catalyst altering the equilibrium point would give a perpetual motion machine. Since such a machine is considered impossible to realise, we again deduce that a catalyst cannot alter the equilibrium position.

The experiments described in this type of argument are purely conceptual and are not actually performed. They may be described as 'thought experiments'. To deduce general scientific results by arguing from conceptual experiments which are not actually performed is an important method of reasoning in several branches of science.

The synthesis and oxidation of ammonia

The reaction between nitrogen and hydrogen to form ammonia

$$N_2 + 3H_2 \rightleftharpoons 2NH_3$$

is reversible and proceeds from left to right with evolution of heat and a reduction in the number of molecules present. The proportion of ammonia in the equilibrium mixture of the three gases is therefore greater at lower temperatures and higher pressures. At a total pressure of 1 atm $(101,325 \text{ N m}^{-2})$ and at room temperature the equilibrium proportion of ammonia is quite large. However, the rate of reaction between nitrogen and hydrogen at room temperature in the absence of a catalyst is so exceedingly slow that even after thousands of years the formation of ammonia by the uncatalysed reaction would be negligible. No really good ammonia synthesis catalyst operating at room temperature has yet been found, although research on this problem is being carried out in a number of laboratories. Early workers who tried to synthesise ammonia were hampered because ideas on chemical equilibrium had not been developed. In 1904, by which time concepts of chemical equilibrium had been formulated, F. Haber began his very important work on the catalytic synthesis of ammonia from hydrogen and nitrogen. He used high pressures to increase the yield of ammonia, and temperatures high enough to give an adequate rate of reaction, but not too high to make the equilibrium yield of ammonia unduly low. In 1908 Haber and R. Le Rossignol published values for the equilibrium constant for the reaction which agreed fairly well with independent measurements made by W. Nernst. A search by Haber and his colleagues for catalysts suitable for operation at high pressures revealed osmium to be very active and uranium to be active. In July 1909 the laboratory apparatus was demonstrated to the industrial chemist C. Bosch. The catalytic synthesis of ammonia is now brought about industrially on a very large scale. The oxidation of ammonia to nitrogen oxides and thence nitric acid leads to ammonium nitrate, of very great importance as a fertiliser. The use of a platinum sponge catalyst for ammonia oxidation was reported by C. F. Kuhlmann to the Scientific Society of Lille in France in 1838, and W. Ostwald in 1901 prepared nitric acid from nitric oxide made by the oxidation of ammonia with a platinum catalyst. The Ostwald process was introduced in the United Kingdom in 1918.

Much work has been carried out on catalysts suitable for ammonia synthesis in industrial plants, and iron catalysts have been developed containing several promoters. A promoter is a substance which when added to a catalyst increases the catalytic activity to a notable extent although the promoter alone does not have any marked catalytic action on the reaction under study. A promoter is often a metal oxide not reducible by hydrogen. Some ideas on how promoters work are gained from a study of solid state chemistry (Chapter 4). The term 'promoter' was used in the early patents of the Badische Anilin und Soda Fabrik and has since been widely used.

Catalysis in organic chemistry

Little work on catalysis in organic chemistry was carried out in the century following the pioneering experiments of Priestley and van Marum. In 1897, P. Sabatier initiated a series of important researches on catalytic hydrogenation of organic substances. The organic substance, as vapour, was passed with excess hydrogen over a heated metal catalyst. Sabatier with J. B. Senderens found that nickel, cobalt, copper and iron acted as hydrogenation catalysts, nickel being most efficient. The metal was prepared in a fine state of division by reduction of the oxide at about 300°C with hydrogen.

Sabatier with his colleagues extended his research and investigated the catalysis of reactions of dehydrogenation, hydration, dehydration, oxidation, halogenation, elimination, condensation, isomerisation, polymerisation, depolymerisation, and decomposition.

Sabatier supposed that in general a catalyst functions by forming an intermediate compound with a reactant, and this intermediate may subsequently react with another reactant, regenerating the catalyst, or decompose, again regenerating the catalyst. This general idea had in fact been proposed in 1806 by C. B. Désormes and N. Clément as an explanation of the action of nitrogen oxides in the lead chamber process for the manufacture of sulphuric acid, and had been subsequently developed by many workers. Sabatier wrote: 'The theory of catalysis by means of intermediate compounds still contains many obscurities and has the fault of leaning frequently on the assumption of hypothetical intermediate products which we have not yet been able to isolate, but it is the only hypothesis that is able to explain catalysis in homogeneous solution and has the merit of applying to all cases.

'As far as I am concerned, this idea of temporary unstable intermediate compounds has been the beacon light that has guided all my work on catalysis; its light may, perhaps, be dimmed by the glare of lights, as yet unsuspected, which will arise in the better explored field of chemical knowledge.'

2 The adsorption of gases by solids

The occurrence of adsorption

When a solid is exposed to a gas the process of adsorption in general occurs. The word 'adsorption' denotes (in this case of a solid and a gas) an accumulation of particles from the gas phase at the surface of the solid. At this surface there is a gas-solid interface, and the accumulation of particles occurs in the interfacial layer at this interface. The accumulated particles at the interface form what is called the 'adsorbed layer'. The adsorbed species is called the 'adsorbate'.

When a gas molecule approaches the surface of a solid it experiences forces attracting it to the surface and these are generally strong enough to hold the molecule to the surface for a period of time. Indeed, the forces may be so strong that the molecule is disrupted into fragments, or even into its constituent atoms, and these fragments or atoms may then become bound to the solid surface. Because this fragmentation (or dissociation) of molecules often occurs when a gas is adsorbed by a solid we have described adsorption as an accumulation of particles rather than an accumulation of molecules. The particles may be molecules or fragments of molecules. If processes of aggregation occur in the interfacial layer the adsorbed particles may be larger than individual molecules. Adsorption with fragmentation or dissociation, however, is more common than adsorption with aggregation.

The term 'adsorption' is also used to describe the process in which molecules accumulate in the interfacial layer and in which fragmentation of molecules under the influence of strong forces from the solid may occur. The reverse process in which particles escape from the interfacial layer is called 'desorption'. Fragments of molecules on a surface will usually combine with other fragments before desorption occurs, so that the particles desorbed are molecules, but at high temperatures the separate fragments can themselves often be desorbed. For example, methyl iodide (iodomethane) decomposing on heated tungsten gives free methyl radicals which evaporate from the interfacial layer and can be detected in the gas phase by mass spectrometric analysis. Another example is provided by oxygen molecules dissociating on heated platinum to give atomic oxygen which again evaporates from the surface. The heated platinum acts as a catalyst so the yield of atomic oxygen cannot exceed that determined by the equilibrium constant for the reversible reaction

$$O_2 \rightleftharpoons O + O$$

and for a high conversion of oxygen molecules into atoms high temperatures and very low pressures are necessary.

A very important concept in the study of adsorption is that of a steady state in which the rate of adsorption is equal to the rate of desorption and an adsorption equilibrium is established. This equilibrium is dynamic, with molecules constantly bombarding the surface of the solid and being adsorbed at a certain rate, exactly equal for the equilibrium state to the rate at which the same molecules are desorbed.

When a solid is introduced into a gas and adsorption occurs, particles accumulate in the adsorbed layer and cause the apparent weight of the solid to increase. If the solid in a vacuum is hung on a delicate balance and a suitable gas or vapour is introduced, adsorption occurs and the apparent weight of the solid increases by an amount equal to the weight of the adsorbed layer. Buoyancy corrections are necessary in this type of experiment. With a porous solid the increase of weight as a result of adsorption of a suitable gas can be measured without great difficulty, because a large area of surface is effective in bringing about adsorption. Weighing experiments of this kind provide one way of studying adsorption.

Adsorption of a gas can also occur at a liquid surface, that is at the gas-liquid interface. At the interface between a solid and a solution (a solid-liquid interface) both solute and solvent may be adsorbed, although in many systems it is the adsorption of solute which is important. In fact generally, in adsorption from solution by a solid, all the components comprising the solution will be present at the phase boundary and observable phenomena will arise from a difference between the composition of the fluid in the interfacial layer and that in the bulk fluid phase.

A condensed phase (i.e. a liquid or solid phase) at the surface of which adsorption occurs is called an 'adsorbent'. When the species which is first adsorbed on a solid surface then penetrates the surface of the solid and enters the bulk of the condensed phase the process is called 'absorption' and the condensed phase is called the 'absorbent'.

The importance of adsorption for heterogeneous catalysis

M. Faraday in his paper of 1834 (see p. 4) on the combination of hydrogen and oxygen on the surface of platinum proposed that the two reacting gases were condensed on the surface of the platinum and 'the approximation of the particles to those of the metals may be very great'. The two reactants were in this way brought within the action of their mutual affinities. Water formed from hydrogen and oxygen was, according to Faraday's proposal, less attracted by the platinum than the reacting gases and so could evaporate. In modern terms, therefore, the catalytic reaction was supposed to occur in the adsorbed layer. According to this view, which is now generally accepted, the phenomena of adsorption and catalysis are closely related.

The adsorption theory of heterogeneous catalysis had been proposed before Faraday's paper by A. Fusinieri in a paper of 1825, which refers to the continual renovation of *concrete laminae* on the surface of platinum when combustion occurs on it as described by H. Davy. Faraday remarked in his paper that he could not form a distinct idea of the power to which Fusinieri referred the phenomenon, but this was partly due, added Faraday disarmingly, to his imperfect knowledge of the language (Italian) of the memoir.

Similarly adsorption is generally considered to be involved when catalytic reaction occurs at the solid-liquid interface. Adsorption at this interface is, however, on the whole a more complex process than at the solid-gas interface. Furthermore, during the past few years it has become possible actually to observe separate atoms at metal surfaces, and certain processes involving individual atoms which occur when a metal is exposed to a gas have been observed directly. The very remarkable experimental methods which have been developed to make such observations are used under very well-defined conditions and are particularly relevant to the study of adsorption and catalysis by metals. These new methods, in which atomic events are observed, can be used only to investigate phenomena at the solid-gas interface. In this Chapter and in Chapters 3 and 4 of this book we therefore consider only phenomena at the solid-gas interface, although we must note at this point the very great importance of catalysis at the solid-liquid interface, in both industrial and laboratory practice.

Physisorption or physical adsorption

The general idea that a solid surface will exert forces on gas molecules so that adsorption occurs was developed by many workers after the time of Faraday. As early as 1839, J. F. Daniell, Professor of Chemistry at King's College, London, discussed 'a force of heterogeneous adhesion' and the bearing of this force on adsorption phenomena. Seventy years later this view was objected to in a paper from University College, London, by Ida F. Homfray who, in discussing some very elegant experimental results she had obtained on the uptake of a number of gases by charcoal, advanced objections to explanations on the basis of chemical combination or surface condensation, and favoured a hypothesis proposing solution of the gas in the solid. At about the same time, however, A. Titoff and S. Arrhenius both noted that, for the adsorption of a variety of gases on porous solids, the amount adsorbed increases with the constant a of the van der Waals equation of state, and hence with the attraction of the gas molecules for each other. Therefore, they reasoned, the underlying cause of adsorption is to be sought in molecular attraction between molecules of the adsorbed substance and the adsorbent. This view is now generally accepted. Actually Daniell in his book of 1839 had noticed the relation between the extent of adsorption of different gases under comparable conditions and the ease of condensation of the gases.

Chemists now call the type of adsorption discussed by Daniell, Titoff and Arrhenius 'physisorption' or 'physical adsorption', which may be defined as adsorption in which the forces involved are intermolecular forces (van der Waals forces) of the same kind as those responsible for the imperfection of real gases and the condensation of vapours, and which do not involve a significant change in the electronic orbital patterns of the species involved. This particular definition was proposed in a tentative form, open for comment and discussion, by the International Union of Pure and Applied Chemistry in 1970. It is clearly necessary that terms used in science in an exact sense should have exact definitions which are internationally agreed, and an important part of the work of various international bodies is the drafting of such definitions. In fact, years of work may lie behind an internationally agreed definition.

The International Union also listed seven features which are useful in recognising physisorption. For the gas-solid interface these are the following:

(1) The physisorption phenomenon is a general one and occurs in any gas/solid system, although there is also the possibility that certain specific molecular interactions may occur, arising from particular geometrical or electronic properties of the adsorbent or of the gas molecules adsorbed or of both.

(2) The experimental evidence for the perturbation of the electronic states of adsorbent and adsorbate is minimal.

(3) The adsorbed species are chemically identical with those in the gas phase so that the chemical nature of the gas is not altered by adsorption and subsequent desorption. Molecules in the gas in fact are adsorbed as the same molecules and are not disrupted by the forces at the surface.

(4) The energy needed to take an adsorbed molecule away from the solid surface into the gas phase is of the same order of magnitude, although it is usually greater than, the energy needed to take the same molecule from a liquid phase composed of similar molecules into the gas phase.

(5) When a molecule approaches the surface it is adsorbed without having to go over an energy barrier. There is no activation energy for adsorption. (The concept of activation energy is discussed in general textbooks and in other books in this series.)

(6) In physical adsorption an equilibrium is established between the adsorbate (or the adsorbed layer) and the gas phase. The extent of physical adsorption increases with increase in gas pressure and usually decreases with increasing temperature.

(7) The number of molecules adsorbed on the surface may be greater than the number which is in direct contact with the surface. Some molecules are then adsorbed on top of the molecules making direct contact with the surface. When all the molecules present on the surface make direct contact with the surface the adsorbed layer is called a monolayer since it is only one molecule thick. A monolayer becomes complete when no more molecules can be adsorbed so that they make direct contact with the surface. In Fig. 1(a) a diagrammatic representation of a complete monolayer is shown. In Fig. 1(b) an incomplete monolayer is shown. The circles represent adsorbed molecules. In Fig. 1(c) molecules are adsorbed on top of those in the first layer. This is known as multilayer adsorption, and is favoured by high pressures and low temperatures. Multilayer formation may involve the building up of a number of layers of adsorbed molecules.

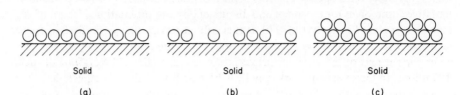

Fig. 1. Monolayers and multilayers: (a) *a completed monolayer;* (b) *an incomplete layer;* (c) *a multilayer has formed*

Chemisorption or chemical adsorption

A series of very important investigations on adsorption was carried out by I. Langmuir in the laboratories of the American General Electric Company. During the period 1912 to 1922, various papers by Langmuir developed a very clear formulation of the concept of chemisorption or chemical adsorption. This concept is now generally accepted. These researches also played an extremely important part in the development of the modern, incandescent, tungsten-filament electric lamp bulb.

During the course of investigations of the heat loss from electrically heated thin wires in gases at low pressures, Langmuir in 1912 developed a theory for the contribution of convection to the heat loss. Measurements of the heat loss from tungsten wires in nitrogen and mercury vapour agreed with the values calculated from this theory at temperatures up to about 3 500 K; but with hydrogen present the heat loss at 3 300 K was four or five times greater than the calculated value. Langmuir discovered that extra heat was taken away from the wire because hydrogen molecules striking the wire could be dissociated into atoms. This reaction is of course endothermic, and the heat needed to bring it about was taken from the tungsten wire. In fact the tungsten acted as a catalyst for the reaction

$$H_2 \rightleftharpoons H + H$$

which is reversible, so that the yield of atomic hydrogen could not exceed that given from equilibrium considerations, just like oxygen dissociating on platinum (p. 8). However, at temperatures as high as 3 300 K the possible conversion into atomic hydrogen at low pressures is quite large. The atomic hydrogen obtained from the wire reacted directly with phosphorus, when it was present in the bulb used, to give phosphine, and showed great chemical activity in other ways.

From this work the concept of a dissociative chemisorption arose. Langmuir supposed that hydrogen molecules approaching the tungsten surface were wrenched asunder near the surface so that atoms of hydrogen became attached to definite atoms or groups of atoms at the metal surface. Definite chemical bonds were supposed to be formed between the hydrogen atoms on the surface and the metal atoms. A diagram of a chemisorbed layer of hydrogen on tungsten is shown in Fig. 2. The adsorption is called dissociative because a molecule of hydrogen dissociates into

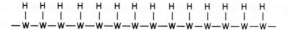

Fig. 2. A monolayer of hydrogen atoms on tungsten

two separate atoms, and it is called chemisorption (or chemical adsorption) because a definite W—H chemical bond is considered to be formed. The adsorbed layer of hydrogen atoms saturates the free valencies of the surface tungsten atoms and thus we expect a layer of adsorbed atoms only one atom thick (that is a monolayer) to be formed in this way.

The International Union of Pure and Applied Chemistry in 1970 defined 'chemisorption' (or 'chemical adsorption') as adsorption in which the forces involved are valency forces of the same kind as those operating in the formation of chemical compounds. The 1970 *Manual of Definitions, Terminology and Symbols in Colloid and Surface Chemistry* of the International Union notes that the problem of distinguishing between chemisorption and physisorption is basically the same as that of distinguishing between chemical and physical interaction in general. No absolutely sharp distinction between the two types of adsorption can be made, and intermediate cases exist. For example, the transfer of an electron from an adsorbed molecule or atom into a solid adsorbent, forming an adsorbed positive ion, or, conversely, the transfer of an electron from the solid adsorbent to an adsorbed molecule or atom, forming an adsorbed negative ion, would be analogous to the formation of an electrovalent chemical bond and we would describe the charged adsorbed layer as chemisorbed. However, if no transfer of an electron occurs and the criteria for physisorption already mentioned are satisfied, we would regard the adsorbed layer as physisorbed. An intermediate possibility is a slight transfer of charge into or out of the adsorbed layer — we could then regard the adsorbed layer as slightly polarised. There would be an electric moment (a dipole moment) at the surface. If the transfer of charge is slight, so that in effect it corresponds to only a fraction of an electronic charge entering or leaving a molecule, then the molecule could be regarded as physically adsorbed. We would refer to 'weak charge transfer' in such a case. Indeed, experiments of several kinds show that physisorbed layers do often have small dipole moments associated with them. Since charge transfer can range from this weak charge transfer to transfer of a whole electronic charge we cannot distinguish sharply between physisorption and chemisorption. However, the two extremes are readily recognisable.

The International Union listed in the 1970 Manual already mentioned seven features which are useful in recognising chemisorption. For the gas-solid interface these are the following:

(1) The phenomenon is characterised by specificity. Whereas physisorption occurs quite generally, for chemisorption one must have both an appropriate solid and an appropriate gas. There is a specific interaction, like chemical reaction in general.

13

(2) Changes in the electronic state may be detectable by suitable physical methods of investigation. Thus, with sufficiently delicate methods, the electronic state of the adsorbed particles can be investigated by ultraviolet, infrared and microwave spectroscopy. Changes in the electronic state of the solid adsorbent can also be investigated in several ways. The entry of electrons into a thin film of metal as a result of adsorption can give rise to an increase in the electrical conductivity of the film, and the withdrawal of electrons from a thin metal film to form a covalent bond with an adsorbed species, or to form a negative ion at the surface, can give rise to an increase in the electrical resistance of the film. Again, unpaired electrons in a solid may be paired, when adsorption occurs, with electrons of opposite spin supplied by the adsorbed particles, and the paramagnetic susceptibility of the solid is decreased in consequence.

(3) The chemical nature of the adsorbate or adsorbates may be altered by surface dissociation or reaction in such a way that on desorption the original species cannot be recovered. In this sense, chemisorption may not be reversible. For example, a mixture of hydrogen and deuterium when adsorbed on tungsten will, according to the model of the adsorbed layer shown in Fig. 2, form a monolayer of adsorbed atoms containing both H and D atoms. On desorption the atoms will combine with neighbouring atoms (unless the temperature is very high so that free atoms evaporate from the surface). Therefore desorption will lead to a mixture of H_2, D_2 and HD. In fact, experiment shows that tungsten will catalyse the reaction of H_2 with D_2 to form HD so efficiently that the reaction can be observed even at the temperature of liquid air.

(4) The energy change involved in chemisorption is of the same order as the energy change in a chemical reaction between a solid and a gas. Chemisorption is usually exothermic with the heat evolved being much greater than the heat evolved by a physisorption. However, just as endothermic chemical compounds can be formed so there is the possibility of realising an endothermic chemisorption under suitable conditions. Generally, however, a spontaneous chemisorption will be exothermic, like the adsorption of hydrogen as atoms on tungsten. We could bring about a similar endothermic chemisorption of a diatomic gas on a metal surface by dissociating the molecules in the gas phase into atoms by some means, for example with an electric discharge, and then allowing the free atoms to strike the metal surface. This, however, would not be a spontaneous chemisorption. Exposure of a metal surface to a gas as molecules does not result in appreciable spontaneous chemisorption of the gas as atoms when the adsorption process is endothermic.

(5) The process of chemisorption often involves an activation energy – an energy barrier exists, which must be surmounted before chemisorption occurs. Generally chemisorption will then proceed slowly and the rate of chemisorption will increase with temperature in the usual way for an activated process. In fact the rate will increase with temperature according to the function $\exp(-E/RT)$ where E is the activation energy per mole, R is the gas constant and T is the thermodynamic tem-

14

perature. (This important exponential function is explained in general textbooks and in other books in this series.)

(6) When the activation energy for adsorption is large, true equilibrium may be achieved only slowly, or in practice not at all. This is because the rate of adsorption becomes so small when E is large.

(7) Since the adsorbed molecules (or particles) are linked to the surface by valency bonds they will usually occupy certain definite adsorption sites on the surface and only one layer of chemisorbed molecules (or particles) is formed. A chemisorbed monolayer is formed.

Adsorption isotherms, isobars, and isosteres

The quantity (x) of a particular gas adsorbed by a certain solid is proportional to the mass (m) of solid present provided the solid has sufficiently uniform properties— for example the range of sizes of any particles composing the solid must be the same from one sample to another so that the surface area of solid per unit mass of solid is the same from one sample to another; and the surface properties must be uniform from one sample to another. Then the quotient x/m measures the extent to which adsorption occurs. When there is a steady state an equilibrium exists between the adsorbed layer and the gas phase, and for any component, which can be denoted generally by A,

$$\text{A (adsorbed)} \rightleftharpoons \text{A (gaseous)}.$$

Generally adsorption is exothermic, so that heat is absorbed when this equilibrium is displaced from left to right. Therefore the equilibrium extent of adsorption (x/m) for exothermic adsorption of A will decrease when the temperature is increased, the gas pressure being kept constant. As the pressure of A is increased, with the temperature kept constant, the extent of adsorption increases because the number of gas molecules colliding with the surface of the solid is directly proportional to the pressure of the gas. Therefore the extent of adsorption x/m is a function of both temperature T and gas pressure p and can be expressed formally as

$$x/m = f(p,T).$$

A complete representation of the extent of adsorption as a function of temperature and pressure is given by setting up three axes at right-angles to represent the three quantities $x/m, p,$ and T. The relation between these three quantities is then given by a surface in space.

In actual practice it is usually easier to consider how any two of the three quantities are related when the third quantity is kept constant. An adsorption isotherm shows the relation between x/m and p, with T kept constant. It is called an isotherm because T is constant. An isobar refers to constant pressure, and shows the relation between x/m and T, with p kept constant. An isostere shows the relation between p and T, with x/m kept constant.

15

The direct observation of adsorption on metal surfaces by field emission microscopy

The field emission microscope provides one of the most direct means of observing the formation of adsorbed layers on a metal surface. This instrument has been developed by E. W. Müller working first in Berlin and later at the Pennsylvania State University, USA. The microscope depends on the emission of electrons from a metal which occurs when a very strong electric field (about 10^9 V m^{-1}) is applied at the metal surface in such a direction that electrons are pulled out of the metal. This emission of electrons is called field emission. Such a field can be produced by shaping a metal wire so that at its end a very fine tip is formed where the curvature of the surface may be so great that it corresponds to a radius of only 100 nm or less. A needle-shaped piece of metal is formed in this way which, when negatively charged, will emit electrons from the very sharp tip. At the tip there is a concentration of electric lines of force and a very strong electric field can be produced. In fact, for a free sphere of radius r at a potential V the field at the surface is V/r. This can be deduced from simple considerations of electrostatics. For a needle-shaped piece of metal, the field at the tip where the radius is r is about $\frac{1}{5}$ of that for a free sphere of the same radius r, for equal potentials applied to each. The field at the tip can be calculated from equations of electrostatics if the tip profile is found with an electron microscope, or it can be found from experiments with a model of the tip in an electrolytic tank. The concept of lines of force is not needed when the field is calculated. In fact, it is not needed when a qualitative explanation is required of why there is a strong electric field at the surface of a charged conductor shaped to have a sharp tip. Instead, the concept of density of electric charge may be used. At the tip there is a high density of electric charge and thus a very strong electric field is produced. The concept of charge density is very useful in the mathematical treatment of electrostatic problems. Both concepts, that of lines of force and that of charge density, have been used in the development of electrostatics.

In the field emission microscope the electric lines of force extend outwards from the tip to a conducting luminescent screen. The general arrangement is shown in Fig. 3. The electrons emitted from the tip are accelerated to the screen along the lines of force, and when they strike the screen it becomes luminous, just like the screen of a cathode-ray tube. The effect is to form an image of the tip of the needle-shaped emitter on the luminescent screen. The microscope operates as a simple projection microscope with electrons as the imaging particles. The presence of an adsorbed layer on a metal surface may greatly alter the field emission of electrons. A positively charged adsorbed layer increases the emission and the image in the microscope becomes brighter. A negatively charged adsorbed layer decreases the emission and the image becomes darker. Thus the formation and some of the properties of adsorbed layers may be seen directly. The movement of a positively charged adsorbent over the metal may be seen as a movement of a bright region on the microscope screen. A single atom of barium on a tungsten emitter will produce a

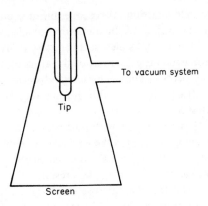

To vacuum system

Tip

Screen

Fig. 3. A schematic diagram of a field emission microscope

visible bright image of itself on the screen because it becomes positively charged by loss of one electron to the tungsten, and therefore greatly enhances field emission. The screen shows a bright spot of light which is the image of an individual barium atom. So experiments with single adsorbed atoms are possible.

There is a cylindrical form of the field emission microscope in which a thin wire is mounted along the axis of a glass tube, the inner surface of which is made a conducting luminescent screen. The thin wire is negatively charged and caused to emit electrons by field emission. Great magnification occurs along a circumference of the glass tube but only small magnification occurs along the length of the tube. So a single adsorbed atom gives a streak as its image. The cylindrical form of microscope can also be used in a different way because very long and thin projections grow readily when a wire is used as a field-emitter in the presence of various gases, and the end of such a projection can be used as an emitting tip in the cylindrical microscope, which then gives images in a manner similar to that of the ordinary field emission microscope. The cylindrical form of microscope is not much used in research, but a simple version can be set up and operated in a modestly equipped laboratory.

The magnification of the field emission microscope is not quite enough to permit observations of single atoms when images of clean metal surfaces, not covered by adsorbed layers, are obtained. Again, the individual atoms in an adsorbed layer on a metal cannot generally be resolved by the microscope, although, as with barium, an image of one single isolated adsorbed atom may be obtainable. The resolution of the field emission microscope is limited to 1·5 to 2·0 nm because the emitted electrons have components of velocity parallel to the surface of the metal.

Field ion microscopy and field ionisation

The field ion microscope, developed by E. W. Müller, has sufficient magnification to produce images of the individual atoms at a metal surface. A needle-shaped emitter is again used, as in the field emission microscope, but it is positively charged. The electric field very near the surface of the metal tip is made so large that atoms, such as those of helium or neon, or molecules, such as those of hydrogen, when near the surface, lose an electron which is pulled out of the atom by the electric field and enters the metal. This process of ionisation in an extremely strong electric field is called field ionisation. The positive ions thus formed (e.g. He^+, Ne^+ or H_2^+) are accelerated to a luminescent screen where an image of the tip surface is formed. The microscope is a projection microscope with positive ions as the imaging particles. The pressure of the imaging gas (that is the gas such as helium or neon or hydrogen which is ionised) must be kept low enough to avoid electrical discharges, and this limits the flux of ions reaching the screen. As a result, field ion microscope images are much dimmer than field emission microscope images. The field ion microscope can give a very high resolution of 200 to 300 pm. The probability of ionisation of the imaging gas atoms or molecules in the field ion microscope depends strongly on the surface topography of the sharp tip on an atomic scale so an image of the tip surface showing many atoms can be produced. Components of velocity of the imaging ions parallel to the metal surface are reduced by using a tip cooled to a low temperature (20 to 70 K) and allowing the atoms or molecules of imaging gas to be cooled by contact with the cold surface before ionisation occurs.

The electric fields used in field ion microscopy are about ten times larger than those used in field emission microscopy. The electric field produces a strong force on the surface at the tip of the needle-shaped emitter, just as there is a force on the plates of a parallel-plate capacitor. This force on the tip can be strong enough to pull off layers of adsorbed atoms or molecules, and in fact even to pull off atoms of the metal itself. This pulling-off process is called field desorption. The field desorption of adsorbed layers has limited the application of field ion microscopy to the study of adsorption phenomena, since it has only been possible to study very strongly bound adsorbed layers, such as those of metal atoms adsorbed on a different underlying metal.

A very recent important advance due to E. W. Müller is his development of the atom-probe field ion microscope. With this technique a single atom of interest viewed on a surface by field ion microscopy is selected for examination. The image of the atom is then caused to coincide with a small probe hole in the image screen. The selected atom is pulled off as an ion by application of a sudden pulse of voltage to the tip, and the ion passes through the probe hole and enters a time-of-flight mass spectrometer which determines its mass/charge and hence its chemical nature. In fact, chemical analysis of a single atom is made. The study of adsorbed layers by this technique, and perhaps even of adsorbed macromolecules, is still in its infancy.

Techniques involving the desorption of positive ions from a surface by the application of a sudden pulse of voltage, and subsequent analysis of the ions by mass spectrometry, have been developed by J. Block in Berlin and by H. D. Beckey in Bonn, Germany, for the study of catalytic decompositions under conditions where

18

there is no electric field during the reaction. Reaction is allowed to occur in an adsorbed layer for a known time, and then the adsorbed layer is field desorbed with a sudden pulse of voltage and the positive ions are analysed mass spectrometrically. In such researches electrical measurements of sufficient delicacy to observe a single positive ion have been made.

Ultra-high vacuum

Researches with metal surfaces using the techniques of field emission, field ionisation, and field desorption have been carried out with really clean metal surfaces. A really clean metal surface is composed of metal atoms which are chemically the same as those of the underlying metal, and the surface is not covered, or only covered to a negligible extent, with an adsorbed layer. Such surfaces can only be prepared and maintained in a very good vacuum, now called an ultra-high vacuum. This term refers to a vacuum system in which the pressure is less than about 10^{-6} N m^{-2}. When the residual pressure is about 10^{-8} N m^{-2}, several hours are needed for residual gases in a vacuum system to form a monolayer on a surface even when every gas molecule striking the surface is adsorbed. This is because the number of collisions of gas molecules with the surface is proportional to gas pressure and becomes very low at these exceedingly low pressures. It is only within the last twenty years that methods of producing ultra-high vacuum have become widely used and made possible many experiments on catalytic action of metal surfaces under such conditions that clean metal surfaces can be produced with the surface actually composed of metal atoms. Under ordinary conditions most metal surfaces become oxidised very quickly and contaminated in many other ways.

3 The rate of heterogeneous catalytic reactions and the adsorption theory

Chemical kinetics and heterogeneous catalysis

The detailed study of the rates of chemical reactions has been pursued with ever increasing vigour over about a century. This field of study is now usually described as 'reaction kinetics'. Reaction kinetics is concerned with the determination of the rate of chemical change and the way in which this rate varies with the pressure or concentration of reactants, and with temperature and other variables. More fundamentally, perhaps, reaction kinetics is also concerned with the interpretation of the observed rate of a given reaction in terms of a detailed mechanism involving molecular events whereby the reaction is supposed to proceed. In fact, a mechanism of reaction is proposed which explains the observed reaction kinetics. The proposed mechanism can be regarded as a hypothetical model system which reproduces the observable behaviour of the real system. The real system may not in fact behave at the molecular level in the manner postulated when the model is constructed. The molecular events postulated for the model cannot generally be observed directly in the real system and inference is necessary in deducing a scheme of molecular events from the observable behaviour of a real reaction. So we expect, and indeed find, many controversies in the field of reaction kinetics, with several mechanisms frequently being advanced, each with its own group of protagonists, to explain the kinetics of a particular chemical reaction.

In heterogeneous reactions of gases on a solid surface as compared with homogeneous reactions of gases there is the additional complicating factor of the intervention of the solid surface on which reaction occurs, and the proposal of detailed mechanisms is in consequence generally subject to greater uncertainty in the heterogeneous case.

In this chapter we briefly review some simple ways in which the observed kinetics of reactions of gases on solid surfaces have been interpreted with the help of the adsorption theory. The discussion is mainly in general terms since the treatment of any given reaction is often still controversial.

The Langmuir adsorption isotherm

The Langmuir adsorption isotherm has played a very important part in the interpretation of the kinetics of heterogeneous catalytic reactions, that is those occurring on a surface on which adsorption occurs.

Consider a solid in equilibrium with a single gas and let part of the surface of the solid be covered by an adsorbed layer of molecules, the remaining part of the surface of the solid being bare (as in Fig. 1(b)). Consider the dynamic equilibrium involving molecules approaching the surface and being adsorbed, and molecules in the adsorbed layer being desorbed into the gas. In a steady state the adsorption rate must equal the desorption rate. A molecule approaching the surface from the gas phase may strike either a bare part of the surface or a part of the surface already covered by adsorbed molecules. I. Langmuir argued that the forces acting between two layers of gas molecules on the surface, one above the other, would usually be very much less than those between the solid surface and the first layer of molecules. This will certainly be so if chemisorption occurs. If it is so, the rate of evaporation in the second layer will be so much more rapid than that in the first that the number of molecules in the second layer will be negligible compared with the number in the first layer. When a gas molecule strikes a portion of the surface already covered it evaporates so quickly that in effect the process is equivalent to a reflection. Therefore to be adsorbed a molecule must, according to this view of Langmuir, strike a bare part of the surface.

Let θ be the fraction of the surface which is covered by adsorbed molecules, so that $1 - \theta$ is the fraction of the surface which is bare, and let p be the pressure of the gas. The number of collisions of gas molecules with unit area of the surface is proportional to p, from the kinetic theory of gases. If this number in unit time is kp, the rate of condensation of gas on the solid surface is $\alpha(1 - \theta)kp$, where α is the probability that a molecule will be adsorbed when it strikes a bare part of the surface. Clearly α must be in the range 0 to 1. If the process of evaporation of an adsorbed molecule has a probability which is not influenced by the presence of other adsorbed molecules, the rate of evaporation of the adsorbed layer (in molecules, or moles, from unit area of surface in unit time) is proportional to θ. Let this rate be $v_1\theta$. We can say that, with this model, the evaporation from the adsorbed layer (that is the desorption) is a process which is kinetically of the first order with respect to the two-dimensional concentration (i.e. molecules, or moles, per unit area of surface) of adsorbed molecules. When the gas is in equilibrium with the adsorbed layer on the solid the adsorption and desorption rates must be equal, so that

$$\alpha(1 - \theta)kp = v_1\theta \qquad\qquad (3.1)$$

whence, if $b = \alpha k/v$

$$\theta = bp/(1 + bp) \qquad\qquad (3.2)$$

21

Equation (3.2) or some variation of it is usually called the Langmuir adsorption isotherm. Since a constant temperature has been assumed for the deduction and since θ is proportional to the quotient x/m (p. 15), equation (3.2) is an isotherm (as defined on p. 15). According to the Langmuir adsorption isotherm, x/m is proportional to the gas pressure at low pressures (when $bp \ll 1$), and at high pressures a limit to adsorption is reached when the surface becomes covered with a monolayer and, very nearly, $\theta = 1$. Then $bp \gg 1$.

Decomposition of a single reactant

A fundamental concept of the adsorption theory of heterogeneous catalysis is that only adsorbed molecules (or particles) can react. The molecules must be adsorbed in order to come under the influence of forces at the surface of the solid. Many reactions are known which involve the decomposition of a single reactant gas in the presence of a solid catalyst. Suppose that reaction only occurs in an adsorbed layer at the surface of the catalyst, and that the number of molecules adsorbed per unit area of this surface is n. If the surface is slightly covered ($bp \ll 1$), n is proportional to reactant gas pressure p. If reaction only occurs in the adsorbed layer we may postulate that the rate of reaction (in molecules, or moles, per unit area of surface) is proportional to n, and hence to p. For reaction in a system of constant volume the kinetic equation is therefore

$$- \mathrm{d}p/\mathrm{d}t = k_1 p \tag{3.3}$$

where t is time and k_1 a constant related to the volume of the reaction vessel, the surface area of the catalyst, and the various terms of the Langmuir isotherm. In contrast, if the surface is largely covered ($bp \gg 1$), n is virtually independent of p, and if the rate of reaction is again proportional to n, the kinetic equation is

$$- \mathrm{d}p/\mathrm{d}t = k_2 \tag{3.4}$$

where the constant k_2 is not dependent on gas pressure p. Equation (3.3) represents a first-order reaction, and equation (3.4) represents a zero-order reaction. In fact, many examples are known of catalytic decompositions of gases on solids which are first-order at low pressures and become zero-order at higher pressures. This is found, for example, for the decomposition of formic acid on a variety of catalysts.

The argument given above leading to equations (3.3) and (3.4) implicitly assumes that the rate of reaction is determined by processes occurring in the adsorbed layer, and not by rates of adsorption and desorption.

The Langmuir-Hinshelwood mechanism

For surface catalytic reaction between two reactants A and B, I. Langmuir proposed a model according to which molecules of A and B (or particles derived from them) must be adsorbed side by side for reaction to occur. If molecules are adsorbed on definite sites, as in chemisorption, then a molecule of A and a molecule of B must be adsorbed on adjacent sites in order to react with each other. This general view was subsequently developed by C. N. Hinshelwood in a series of researches with his students, and this kind of mechanism involving adjacent adsorption is now called a Langmuir-Hinshelwood mechanism. If the rate of reaction is again determined by processes occurring in the adsorbed layer, and not by rates of adsorption and of desorption, it would seem probable that the rate should be proportional to $\theta_A \theta_B$, where θ_A is the fraction of the surface covered by substance A and θ_B that covered by substance B. For two gases A and B in contact with a surface the steady-state conditions give two equations exactly analogous to equation (3.1). These equations are

$$\alpha_A \left(1 - \theta_A - \theta_B\right) k_A p_A = v_A \theta_A \tag{3.5}$$

$$\alpha_B \left(1 - \theta_A - \theta_B\right) k_B p_B = v_B \theta_B \tag{3.6}$$

where p_A and p_B are the pressures of A and B. If both gases are only slightly adsorbed, $\theta_A \ll 1$ and $\theta_B \ll 1$, therefore, very nearly,

$$\theta_A = C_A p_A$$

$$\theta_B = C_B p_B$$

where C_A and C_B are constants. If the rate of reaction R is

$$R = Q \theta_A \theta_B \tag{3.7}$$

where Q is a constant, then

$$R = Q C_A C_B p_A p_B. \tag{3.8}$$

In fact if both reactants are weakly adsorbed the reaction kinetics are similar to those of a second-order gaseous reaction.

Another simple case arises when one reactant (say A) is weakly adsorbed and the other B is strongly adsorbed, so that $\theta_B \approx 1$. It is supposed that the weak adsorption of A occurs on the part of the surface not covered by B. This part is of fractional extent $1 - \theta_B$. Hence

$$\theta_A \ll 1 - \theta_B$$

and equation (3.5) becomes approximately

$$\theta_A = C_A \left(1 - \theta_B\right) p_A. \tag{3.9}$$

Since $\theta_B \approx 1$ the rate expression (3.7) becomes

$$R = Q C_A \left(1 - \theta_B\right) p_A. \tag{3.10}$$

23

Equation (3.6) for the adsorption of B becomes approximately

$$1 - \theta_B = 1/C_B p_B \qquad (3.11)$$

and equation (3.10) therefore gives

$$R = (QC_A/C_B)(p_A/p_B). \qquad (3.12)$$

The rate of reaction according to this model is inversely proportional to the pressure of strongly adsorbed gas, p_B. This arises because it follows from the Langmuir isotherm that the fraction of surface not covered by a strongly adsorbed gas is inversely proportional to the pressure of that gas, as shown in equation (3.11).

Before the formulation of the Langmuir theory, M. Bodenstein and C. G. Fink in 1907 had found that the rate of oxidation of sulphur dioxide to sulphur trioxide on various solid catalysts was decreased by an increase of sulphur trioxide pressure. They had proposed a model according to which the sulphur dioxide diffused through a layer of sulphur trioxide to reach the catalytic surface. This diffusion was supposed to determine the rate of reaction. As the sulphur trioxide layer increased in thickness with increasing sulphur trioxide pressure, the rate of reaction decreased. An alternative view advanced later by Langmuir was that sulphur trioxide is strongly adsorbed. Then the rate of reaction falls with increase of sulphur trioxide pressure as the surface area of catalyst available for reaction is reduced. In fact a strongly adsorbed product can reduce the rate of reaction just as a strongly adsorbed reactant can. The view of Langmuir is now generally accepted.

When the kinetics of a surface catalysed reaction are deduced with an equation of the type of (3.7) it is implicitly assumed that the observed rate of reaction is determined by processes occurring in the adsorbed layer and not by rates of adsorption or desorption.

It is usually assumed in proposing a Langmuir-Hinshelwood type mechanism for any particular reaction that both reactants must be chemisorbed, so that they are both influenced by strong forces from the surface.

The Rideal mechanism

E. K. Rideal in 1939 drew attention to a difficulty presented by the adjacent interaction (Langmuir-Hinshelwood) mechanism. This arises if both reactants are strongly chemisorbed to the surface, when strong chemical bonds presumably exist between surface sites and molecules or fragments of molecules on the surface. For reaction to occur these strong bonds must be broken, and it is a problem to see how this can occur rapidly. Rideal therefore proposed that surface catalytic reactions take place between a chemisorbed radical or atom and a molecule which comes directly from the gas phase or from a physically adsorbed layer of molecules adsorbed on top of the chemisorbed layer. This physically adsorbed layer is often called a van der Waals layer because the forces holding the molecules to the surface are similar to those

24

giving rise to the term a/V^2 in the van der Waals equation of state. Rideal further proposed that the interaction mechanism is such that the strong bond between the chemisorbed constituent and the catalyst is either not broken, or alternatively, a new chemisorbed bond is formed as the old bond is broken. Thus energy is provided to break the old bond. As an example, Rideal considered the reaction of ethylene chemisorbed on nickel with deuterium held by van der Waals forces. The mechanism he proposed is shown below; a chemisorbed bond is denoted by a continuous line and a van der Waals adsorption by a dashed line:

$$
\begin{array}{ccc}
CH_2{-}CH_2 & D_2 \\
| \quad\quad | & \vdots \\
Ni \quad Ni & Ni
\end{array}
\longrightarrow
\begin{array}{ccc}
CH_2{-}CH_2D{-}D \\
| \quad\quad\quad \vdots \quad\quad \vdots \\
Ni \quad Ni \quad Ni
\end{array}
$$

$$
\longrightarrow
\begin{array}{ccc}
CH_2{-}CH_2D & D \\
| \quad\quad \vdots & \vdots \\
Ni \quad Ni & Ni
\end{array}
$$

Here, as a strong Ni—C bond is broken, another strong bond between Ni and D is formed.

Although the Rideal mechanism envisages as one possibility reaction between a chemisorbed particle and a molecule which strikes that particle from the gas phase, the fundamental concept of the adsorption theory of catalysis that only adsorbed molecules (or particles) can react is in fact retained. The molecule approaching from the gas experiences forces as it approaches the surface and it must become physisorbed, although this physisorption may be of only short duration. The original model of Langmuir for adsorption equilibrium in the absence of reaction proposes that a molecule striking a covered part of a surface evaporates so quickly, that in effect the process is equivalent to a reflection, but this is equivalent to saying that a weak physisorption of short duration occurs. Detailed experimental studies of the behaviour of molecules striking surfaces on which they are not chemisorbed show that a true reflection of a beam of incident molecules from a surface, with the angle of incidence equal to that of reflection, is a rare phenomenon. Generally a physisorption occurs although it may be of very brief duration.

4 Catalysis and solid state chemistry

Deviations from stoicheiometric composition in ionic crystals

Many catalysts of practical significance consist of ionic crystals, in which the units from which the structure is built are ions. One way in which catalytic power has been related to the chemistry of the solid state has been through the study of non-stoicheiometric solids as catalysts.

Consider an ionic crystal composed of ions M^{2+} and X^{2-}, where M represents a metal and X represents an electronegative element, such as oxygen or sulphur. We normally regard the formula of the binary compound MX as being accurately represented by MX. If this is so the compound may be described as stoicheiometric. Many compounds are known in which one of the components M or X is in slight excess, so that the formula may be written MX_{1-x} or MX_{1+x} where $x \ll 1$. The formula could, of course, just as well be written $M_{1-x}X$ or $M_{1+x}X$. The compound is stoicheiometric in the limiting case when $x = 0$.

When the formula is MX_{1-x} the component M is present in slight excess. This excess can arise in two ways. First, it is possible that some extra metal ions are squeezed in at interstitial positions in the crystal structure. These interstitial positions are places which would not normally be occupied by ions in a stoicheiometric crystal. A diagrammatic representation of an interstitial M^{2+} ion is shown in Fig. 4. A circle has been drawn round the interstitial ion to emphasise its position. An important result follows from the requirement that the crystal as a whole must be electrically neutral. The extra M^{2+} ion is introduced actually as an atom, which

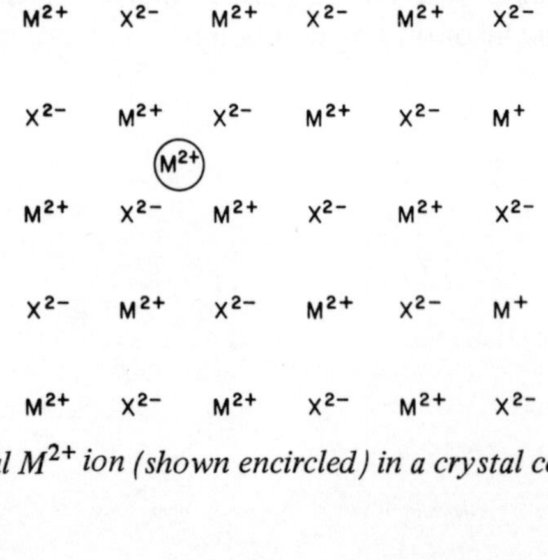

Fig. 4. *An interstitial M^{2+} ion (shown encircled) in a crystal composed of M^{2+} and X^{2-} ions*

ionises giving two electrons, and these electrons must also be present in the crystal. In Fig. 4 these two electrons are supposed to have changed two M^{2+} ions in normal positions in the structure into M^+ ions. The whole structure therefore contains exactly equal numbers of positive and negative charges. If the interstitial metal ion was introduced into the crystal as an ion, without the compensating electrons, then the crystal would become positively charged. But this is against experience. One way of actually making a non-stoicheiometric crystal of the type shown in Fig. 4 might be by heating a stoicheiometric MX crystal in the vapour of the metal M when atoms of M could diffuse from the surface of the crystal into the interior, by moving, as ions, from one interstitial position to another. It is atoms which enter the crystal and so electrical neutrality is preserved. The atoms ionise at the surface or in the interior forming positive ions M^{2+} each of which must be accompanied by two electrons. In Fig. 4 these electrons are attached to the ions shown as M^+ rather than as the normal M^{2+}. This is a somewhat qualitative picture – a more exact description can be given in very mathematical terms which are not readily interpreted (if at all) in terms of a simple physical picture. When the non-stoicheiometric crystal is composed of uni-valent elements (for example, an alkali metal halide) an electron accompanying an interstitial M^+ ion can be regarded, on this qualitative view, as converting an M^+ ion in the lattice into an atom. Again this is not an exact description.

The second way in which component M can be present in slight excess is for there to be a slight deficit of the other component X. This arises when at some of the places where X^{2-} should be present in the structure the X^{2-} ion is missing and there is only the empty space where the X^{2-} ion should be. This space is described as a lattice vacancy, or more briefly just as a vacancy. A missing X^{2-} ion can be described as a vacancy on an anion site. Again, when such vacancies exist, the crystal as a whole is electrically neutral and therefore each vacant anion site must be accom-panied (in this case where X forms a doubly charged ion) by two electrons, which as before can be visualised qualitatively as two M^+ ions. In Fig. 5 a lattice vacancy is shown with the two accompanying electrons.

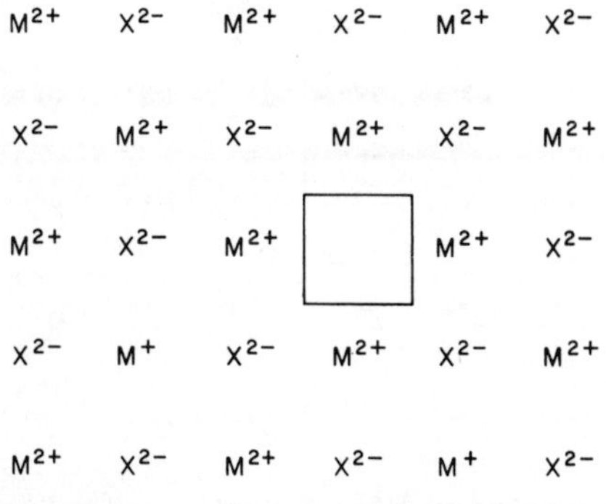

Fig. 5. A lattice vacancy where an X^{2-} ion is missing. The vacancy has been emphasised by drawing a square in it

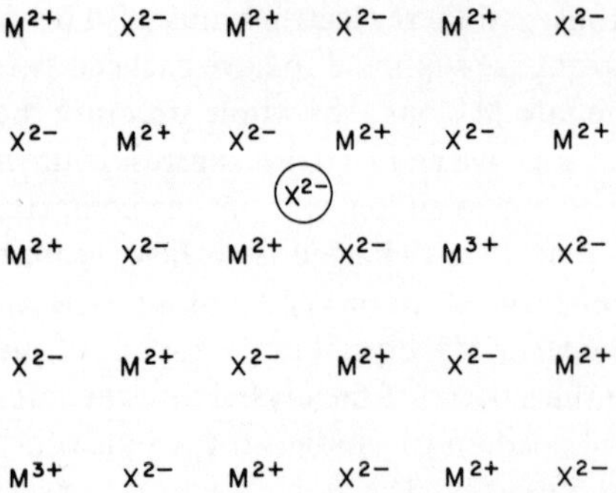

$$M^{2+} \quad X^{2-} \quad M^{2+} \quad X^{2-} \quad M^{2+} \quad X^{2-}$$

$$X^{2-} \quad M^{2+} \quad X^{2-} \quad M^{2+} \quad X^{2-} \quad M^{2+}$$

$$\widehat{(X^{2-})}$$

$$M^{2+} \quad X^{2-} \quad M^{2+} \quad X^{2-} \quad M^{3+} \quad X^{2-}$$

$$X^{2-} \quad M^{2+} \quad X^{2-} \quad M^{2+} \quad X^{2-} \quad M^{2+}$$

$$M^{3+} \quad X^{2-} \quad M^{2+} \quad X^{2-} \quad M^{2+} \quad X^{2-}$$

Fig. 6. An interstitial X^{2-} ion (shown encircled) in a crystal composed of M^{2+} and X^{2-} ions

When the formula is MX_{1+x} (with $x \ll 1$) the component X is present in slight excess. This excess can again arise in two ways: there may be interstitial X^{2-} ions, or there may be lattice vacancies at sites normally occupied by M^{2+} ions. Fig. 6 shows an interstitial X^{2-} ion, again encircled to emphasise its position. The crystal as a whole must be electrically neutral and therefore the two electrons associated with the interstitial X^{2-} ion must be taken from the crystal; in Fig. 6 two of the ions which are normally M^{2+} have had an electron removed from them, leaving two M^{3+} ions. In the whole crystal the numbers of negative and of positive charges exactly balance. The representation of a missing electron as an ion in the lattice carrying an extra positive charge is again a qualitative way of expressing a concept which really needs exact mathematical formulation in terms of equations having no simple physical significance.

Fig. 7 shows the other way in which the component X can be present in slight

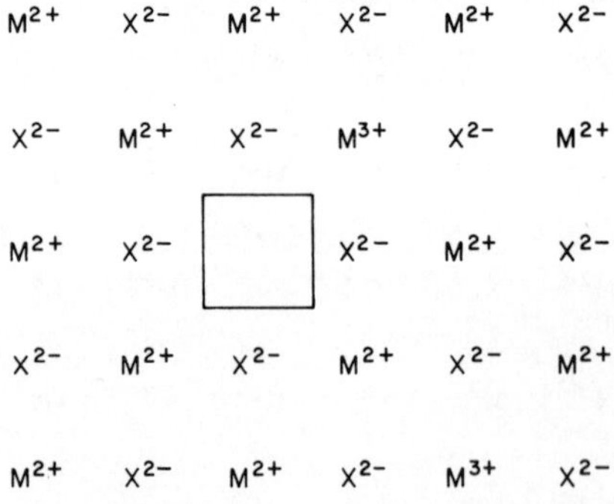

$$M^{2+} \quad X^{2-} \quad M^{2+} \quad X^{2-} \quad M^{2+} \quad X^{2-}$$

$$X^{2-} \quad M^{2+} \quad X^{2-} \quad M^{3+} \quad X^{2-} \quad M^{2+}$$

$$M^{2+} \quad X^{2-} \quad \square \quad X^{2-} \quad M^{2+} \quad X^{2-}$$

$$X^{2-} \quad M^{2+} \quad X^{2-} \quad M^{2+} \quad X^{2-} \quad M^{2+}$$

$$M^{2+} \quad X^{2-} \quad M^{2+} \quad X^{2-} \quad M^{3+} \quad X^{2-}$$

Fig. 7. A lattice vacancy where an M^{2+} ion is missing. The vacancy has been emphasised by drawing a square round it

excess. The figure shows a vacant lattice site at a position where there would normally be an M^{2+} ion. So there is a vacancy on a cation site. Again, to keep the crystal as a whole electrically neutral two ions which are normally M^{2+} have been changed into M^{3+} ions.

A structural imperfection which involves a single atom or ion or a single lattice site is called a 'point defect'. There are other kinds of imperfections in solids which involve large numbers of atoms or ions. For example, the structural imperfection associated with a dislocation extends along a whole line of atoms, as described in Chapter 5.

Another way in which a non-stoicheiometric solid may arise is from substitutional solid solution. Here an atom of one type occupies a place in the structure which is normally occupied by an atom of another type. For example, in a binary compound AB the proper arrangement of atoms along some appropriate line drawn through the structure might be

$$\ldots ABABABABABAB \ldots$$

and if some of the B atoms are replaced by A atoms giving, for example

$$\ldots ABABABAAABAB \ldots$$

a non-stoicheiometric solid is produced containing a slight excess of the component A. This process of substitutional solid solution is not very probable in an ionic crystal. In such a crystal every ion normally has, as its nearest neighbours, ions of the opposite electrical charge. For example, in Figs. 4, 5, 6 and 7 every X^{2-} ion is surrounded by four M^{2+} ions in the same plane, and every M^{2+} ion is surrounded by four X^{2-} ions in the same plane, in any part of the crystal where there is no point defect. Since the concentration of point defects is usually small, the structure of most of the crystal is the normal one without point defects. There will also be neighbours of opposite sign in planes above and below that shown in these figures. When an M^{2+} ion in a normal lattice position is replaced by an X^{2-} ion, this particular X^{2-} ion is surrounded by ions having the same negative charge as itself, and very strong electrostatic repulsions result. Because of this, substitutional solid solution in an ionic solid is not probable. However, substitutional solid solution occurs very widely in alloys and intermetallic compounds. When two substances A and B form an alloy or intermetallic compound, substitutional solid solution involving the replacement of an atom of A by an atom of B puts the B atom in an environment where it has other B atoms as immediate neighbours rather than A atoms. However, when both A and B are metallic in nature the change of environment resulting from substitutional solid solution is not great, and so such solid solution may occur very readily.

One of the earliest laws of chemistry is the law of fixed (or constant) proportions, first clearly stated by J. L. Proust in 1797, although before that time this law had been tacitly accepted by a number of chemists. According to this law, elements combine in definite ratios by weight and the composition of a pure chemical compound is independent of the way in which it is prepared. Proust supported this view by presenting the results of a long series of investigations on the composition of minerals and of compounds of many metals. An alternative view to that of Proust was advanced by C. L. Berthollet in 1803, who contended that the composition of compounds can vary within limits. A controversy occurred between Proust and Berthollet and by about 1807 most chemists had adopted the view of Proust, which gained strong support from the proposal by J. Dalton in 1807 of the atomic theory. Dalton postulated that every kind of atom has a definite atomic weight and that atoms combine to form chemical compounds in definite ratios of whole numbers which are usually small. According to these two postulates the gravimetric composition of a chemical compound is constant and the compound is stoicheiometric.

It is now known that many oxides, sulphides, halides and other compounds of metals can be made slightly non-stoicheiometric. For example, heating oxides, sulphides and halides in oxygen, sulphur vapour or the halogen vapour respectively often produces a crystal containing an excess of the electronegative constituent, and, conversely, heating compounds of these types in a continuously pumped vacuum system often leads to a loss of oxygen, sulphur or halogen, producing a crystal containing an excess of the electropositive component. Refined methods may be necessary to detect the deviations from stoicheiometry, and, of course, such methods had not been developed at the time of Proust and Berthollet.

Non-stoicheiometric ionic crystals as semiconductors

The presence of point defects in an ionic crystal may have a very important effect on the electronic properties of the solid, and in particular a non-stoicheiometric ionic crystal may behave as a semiconductor. A semiconductor has an electrical conductivity which is intermediate between that of a metal and an insulator, and which is not electrolytic, so that no (or negligible) transport of matter occurs during the conduction process. Generally the conductivity of a semiconductor increases strongly with temperature, in contrast to metals which have a resistivity increasing with temperature.

One way in which semiconducting properties can arise may be appreciated from a scrutiny of Fig. 4. This shows a non-stoicheiometric ionic compound containing an excess of the electropositive component, present as interstitial cations. Accompanying each interstitial cation there are two electrons so that the whole crystal is electrically neutral. Generally in an ionic crystal without point imperfections there is a local neutralisation of electric charge — the ions are so arranged with respect to each other that over a distance of a few inter-ionic spacings the positive and negative charges neutralise each other. But when an interstitial positively charged ion is present, as in Fig. 4, the ion creates a local region of positive charge in the crystal. Electrostatic attraction occurs between the interstitial ion and the two electrons associated with it. We can regard the electrons as circling round the interstitial ion —

in fact the situation in some ways resembles that in the helium atom. As the electrons circulate round the ion, different M^{2+} ions in the structure become M^+. In fact the position in the diagram where M^{2+} becomes M^+ can be regarded as continuously changing as an electron circulates. For a single electron circulating round a singly charged interstitial positive ion the situation is in some ways like that in the hydrogen atom.

The energy needed to remove an electron circulating round a proton, thus forming a hydrogen atom, from its lowest energy state in the hydrogen atom to an infinite distance away from the proton can be calculated from the Bohr theory of the hydrogen atom, or by solving the Schrödinger equation for the hydrogen atom. This energy is the ionisation energy or the ionisation potential. The simple Bohr theory adopts a model which equates the centrifugal force on an electron moving round a proton in a circular orbit with the electrical force of attraction between the proton and the electron as given by Coulomb's law. The orbital angular momentum of the electron is then supposed in the simple Bohr theory to be quantised. From these assumptions a value of the ionisation energy of the hydrogen atom can be calculated, and this value agrees very well with the experimental value. An electron circulating round an interstitial positive ion in a crystal is in some ways similar to a hydrogen atom. One very important difference is that the crystal has a relative permittivity (dielectric constant) exceeding unity. The value for inorganic crystals is frequently in the range 5 to 10. A possible very simple model of an electron circulating round an interstitial ion in a crystal would be one similar to the Bohr model of the hydrogen atom with the additional assumption that the two charges are immersed in a uniform dielectric medium of relative permittivity ϵ_r, when the force between the two charges is reduced by a factor ϵ_r compared to the force between the charges in vacuum (the distance between the charges being the same). The effect on the calculated ionisation energy of immersing the charges in a uniform dielectric medium is to reduce it by $1/\epsilon_r^2$. The ionisation energy of a hypothetical hydrogen atom immersed in a uniform dielectric is less by a factor of $1/\epsilon_r^2$ than that of a hydrogen atom in free space, according to calculations based on the simple Bohr model.

In fact, the electrons associated with an interstitial positive ion are only weakly bound to it and can escape from the vicinity of the ion. The simple argument above from the Bohr theory is a way of seeing why this escape occurs readily. Once electrons have escaped, they can move under the influence of an external electric field applied to the crystal and give rise to an electrical conductivity. No transport of matter occurs when the conduction is due only to the motion of electrons. The conduction can be described as electronic. There is also the possibility that the interstitial ion can move slowly under the influence of an electrical field, by jumping from one interstitial position to another. This mode of conduction involves an actual transport of matter through the crystal, and it may be described as ionic or electrolytic. Both the ionic and electronic conductivities of a non-stoicheiometric crystal containing interstitial positive ions increase strongly with temperature. The escape of electrons from an interstitial ion can be regarded as a process which requires an activation energy, and the movement of an interstitial ion from one interstitial position to a neighbouring interstitial position also requires an activation energy, as the interstitial ion has to squeeze past other ions in the lattice. Generally a process requiring an activation energy (an activated process) becomes more probable as temperature increases.

One may say that interstitial cations in a non-stoicheiometric ionic crystal can donate electrons to the lattice as a whole, and these electrons give rise to semiconducting properties. The interstitial cations play a part similar to that of donor impurities in another type of semiconductor, called an impurity semiconductor. An example of a donor impurity is arsenic dissolved in germanium by substitutional solid solution. Germanium has a diamond-type lattice with every germanium atom surrounded by four other germanium atoms tetrahedrally disposed. A very small quantity of arsenic dissolved in germanium produces a huge increase in electrical conductivity. An arsenic atom can replace a germanium atom in the structure. The arsenic atom is bonded to four germanium atoms, and one valency electron from arsenic is unused. This electron can escape into the lattice, leaving behind a positively charged arsenic atom. The electron can then migrate through the lattice and give rise to semiconducting properties.

Fig. 5 represents a vacant anion site and two electrons associated with it. Because a negative ion is missing at the vacant site, electrical charge is not neutralised locally in the region very near the vacancy. There is clearly an excess of positive charge in this region, and this attracts the two electrons. We can imagine the electrons to be circulating around the vacancy, just as they circulate round an interstitial positive ion. Again, the electrons may be able to escape from the influence of the vacancy and then to wander freely through the lattice. If an external electric field is applied, electrons which have escaped from the immediate region of a vacancy can give rise to semiconducting properties. Therefore an anion vacancy can donate electrons to the crystal as a whole. The vacancy behaves like a donor impurity.

A semiconductor in which the current is carried by electrons is called an n-type semiconductor. Various experiments do in fact show that the current-carrying particles behave as electrons.

Semiconducting properties can also arise in a non-stoicheiometric ionic crystal which contains an excess of the electronegative constituent, as in Figs. 6 and 7. In Fig. 6 an interstitial negative ion is shown, and associated with it are two metal ions, each of which has lost an electron, becoming M^{3+} instead of the normal M^{2+}. Therefore each M^{3+} represents a place where an electron is missing. A missing electron is often referred to as an 'electron hole' or, more briefly, simply as a 'hole'. The vacant position left behind when an electron is taken away from an atom or ion can also be described as an empty electron state. There is in fact a state which can accommodate an electron, but in that state there is no electron. A missing electron corresponds to a positive charge, and therefore the vacant position is also called a 'positive hole'.

The terms 'electron hole', 'hole', and 'positive hole' have all been used with the same meaning. It is, perhaps, unfortunate from the point of view of exact terminology, that an 'electron hole' is the same as a 'positive hole'. The term 'hole' refers to something missing, and so an 'electron hole' is, logically, a missing electron. We shall therefore use the terms 'electron hole' or 'hole'.

An electron hole is mobile just as an electron is. Thus in Fig. 6 the two M^{3+} ions are near a number of M^{2+} ions. An electron may transfer from one of these M^{2+} ions to the original M^{3+} ion, which then becomes M^{2+}. But the ion from which the electron has transferred becomes M^{3+}. In effect the electron hole has moved. We can regard the process as involving continual jumps of an electron into the hole. The electron hole is a region of positive charge and so it is attracted electrostatically by the interstitial negative ion. But just as electrons can escape from the influence of an interstitial ion or vacancy, so can electron holes escape from an interstitial ion or vacancy. When they have escaped, the holes can wander through the crystal and migrate under the influence of an applied electric field, giving rise to semiconducting properties.

A semiconductor in which the current is carried by electron holes is called a p-type semiconductor. The current is due to successive jumps of valence electrons into empty electron states. In some respects, however, a p-type semiconductor behaves as though the current carrying entities are actually positively charged.

In Fig. 7 a vacant cation site is shown which has associated with it two electron holes. The vacant cation site corresponds to a local region of negative electric charge in the crystal. Therefore an electrostatic attraction occurs between the cation vacancy and the electron holes. We may say that we now have a situation where something which is missing is interacting with something else which is also missing. The electron holes can escape from the influence of the vacancy to produce p-type semiconduction.

Electron holes can also be produced by impurities introduced into a crystal by substitutional solid solution. An example is provided by boron dissolved in germanium. A boron atom can replace a germanium atom, when it has four germanium atoms around it, and to form four covalent bonds with these four atoms the boron atom needs to take one electron from the crystal as a whole. The boron atom in fact accepts an electron, and an electron hole is left behind. The boron is described as an acceptor impurity. The impurity atom is negatively charged and interacts electrostatically with an electron hole, which can be regarded as a positively charged germanium atom. However, the hole can escape from the impurity atom and produce p-type semiconduction. Interstitial negative ions and cation vacancies can both behave like acceptor impurities.

The investigation of imperfections in ionic crystals by measurement of electrical conductivity

Generally when semiconducting properties arise as the result of the non-stoicheiometric composition of an ionic crystal, the conductivity is due to the motion of electrons or of electron holes. There is, in addition, the possibility of some ionic or electrolytic conductivity, especially at high temperatures. Thus, interstitial ions can move through the lattice under the influence of an applied electric field, and vacancies can also move through the lattice. A vacancy moves when a neighbouring ion of the correct electrical sign enters the vacancy, which then, of course, moves to the original position of the neighbouring ion. Since a vacancy corresponds to a region of local electrical charge it can migrate in an applied electric field by a process involving continual jumps of neighbouring ions into the vacancy.

One way of investigating the type of point defect present in an ionic crystal is by measurements of changes of electrical conductivity of the crystal when it is heated in the vapour of the electronegative constituent. Consider the four different possibilities represented by Figs. 4, 5, 6, and 7. A crystal containing an excess of cations, present as interstitial cations, as in Fig. 4, will, in the most general case, show two types of conductivity: n-type semiconduction due to motion of electrons, and, at high enough temperatures, ionic conduction due to the movement of the interstitial cations through the crystal. The two types of conduction can be distinguished experimentally because actual transport of matter is only associated with movement of the positive ions, which will form the metal M when they reach an external electrode. The semiconduction can be recognised as n-type because the current-carrying particles are electrons, and they experience a slight deflection when a magnetic field is applied to the crystal in a direction at right-angles to the direction of flow of current. This deflection is in the same direction as that of an electron moving at right-angles to a magnetic field. The effect of the deflection in the crystal is to develop a transverse electric field which is at right-angles to both the applied magnetic field and the direction of current flow in the crystal (that is, the direction in the absence of the magnetic field). The development of this transverse electric field is known as the Hall effect. Fig. 8 shows the production of the transverse field E by the Hall effect.

Consider the crystal of formula MX_{1-x} to be heated in the vapour of the electronegative constituent X. Suppose that point defects of the type shown in Fig. 4 are present. Interstitial M^{2+} ions near the surface can move to it and combine there with adsorbed X particles to form more of the crystal. Each X as it is incorporated into the crystal must become X^{2-} and so two electrons are taken from the crystal as a whole. This means that two ions in the crystal change from M^+ to M^{2+}. Each X becoming X^{2-} is accompanied by one M^{2+} ion which ceases to be an interstitial ion and becomes an ion in a normal position in the lattice. The two electrons originally associated with the interstitial ion convert X to X^{2-}. Electrical neutrality is preserved. The reaction occurring can be written as:

$$M^{2+} \text{ (interstitial)} + X \text{ (adsorbed)} + 2 \text{ electrons} \rightarrow M^{2+}X^{2-} \text{ (lattice)}$$

The effect of this reaction is to cause a reduction in the electrical conductivity of the crystal. The number of electrons capable of becoming current carriers is reduced

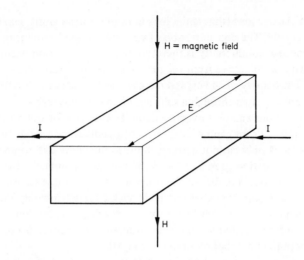

H = magnetic field

E

I   I

H

Fig. 8. Production of a transverse electric field E across a semiconducting block by the Hall effect. The current I flows in the direction shown when H = 0

as the number of interstitial ions is reduced. The electronic and the ionic conductivities are both decreased. Eventually all the interstitial ions may diffuse to the surface and form new crystal. This can occur because the initial movement of interstitial ions near the surface to the surface, followed by their reaction there, produces a concentration gradient of interstitial ions, and hence diffusion of these ions to the surface will continue. When all the interstitial ions have gone, the crystal is stoicheiometric, and the semiconduction and ionic conduction, produced as discussed above, will both disappear.

Another effect of the removal of interstitials is to decrease the density of the crystal. The density of a crystal of the type shown in Fig. 4 is slightly greater than the density calculated for a perfect crystal of the same structure (but without point defects) from the atomic weights of M and X and the lattice spacings in the crystal. In summary, therefore, it can be seen that there are several observable effects which can occur when a crystal of the type shown in Fig. 4 is heated in the vapour of the electronegative component X.

When a crystal with imperfections of the type shown in Fig. 5 is heated in the vapour of the electronegative constituent, anion vacancies can diffuse to the surface where they are filled by reaction with adsorbed X:

$$\text{anion vacancy} + \text{X (adsorbed)} + 2 \text{ electrons} \rightarrow X^{2-} \text{ (lattice)}$$

The vacancies are thereby filled with X^{2-} ions at the surface. Further diffusion of the vacancies towards the surface under the influence of the concentration gradient of vacancies produced as a result of the reaction at the surface is equivalent to a diffusion of X^{2-} lattice ions (that is the ions in normal lattice positions) into the interior of the crystal. The concentration of vacancies in the crystal is reduced. The electronic n-type semiconductivity is therefore reduced as reaction occurs, and the ionic conductivity, which arises from the movement of the vacancies, is correspondingly reduced. The density of the crystal increases as the vacancies are filled.

When a current is passed through a solution of an electrolyte, part of the current is carried by the cations and part by the anions. The concept of transport number is used. Similarly this concept can be used for ionic conduction in solids. For the situation shown in Fig. 4 only cations carry current, so they have a transport number of 1, whereas for the situation shown in Fig. 5 the transport number of the anions is 1. In each case the transport number of the other ion is zero. These transport numbers can be measured with suitable experiments. In such experiments only that part of the current which is carried by the electrolytic part of the conduction is considered. The part of the current carried by the semiconduction mechanism is not considered in finding the transport number of an ion in a crystal.

The heating in the vapour of the electronegative component of a crystal of the type shown in Fig. 6 causes more interstitial anions to enter the crystal. Each extra interstitial anion is accompanied by two electron holes. The heating therefore brings about an increase in the ionic conductivity, which is due to the movement of interstitial anions, and an increase in the p-type semiconductivity. The transport of the anions is 1. The heating also causes an increase of density.

For a crystal of the type shown in Fig. 7, heating in the vapour of the electronegative component causes an increase in the number of cation vacancies. More $M^{2+}X^{2-}$ is formed at the surface of the crystal as M^{2+} ions diffuse to the surface by entering lattice vacancies diffusing away from the surface. The heating increases the ionic conductivity, and the transport number of the cations is 1. The heating also increases the p-type semiconductivity and reduces the density.

When the electronic part of the conductivity is due to the movement of electron holes the crystal behaves when it produces the Hall effect as though the current-carrying particles are actually positively charged. An electron hole behaves with respect to the Hall effect like a positive charge. This means that the sign of the transverse field produced in the Hall effect for an n-type semiconductor is opposite to that for a p-type semiconductor. This is one way of distinguishing between the two types.

The four possibilities of structural disorder represented by Figs. 4, 5, 6 and 7 all lead to different observable effects when the crystal is heated in the vapour of the electronegative component. Similarly, the effects of heating in the vapour of the electropositive component could be deduced, and are again different for the four possibilities. It is important to note that the four possibilities discussed are limiting cases, and in a real crystal several or even all of the possible types of disorder may exist together. For example, vacant lattice points may be just balanced by interstitial particles of the correct charge, so that the density is unchanged. Electrical neutrality of the crystal can then be preserved without the existence of semiconducting electrons and electron holes. In some semiconductors both electrons and electron holes are present and contribute to the semiconductivity.

Semiconductors containing incorporated foreign ions

An ionic crystal may become semiconducting when some of the ions in normal lattice positions are replaced by foreign ions of different valency which occupy normal lattice positions. For example, the Ni^{2+} ion in nickel oxide can be replaced by Li^+. Because electrical neutrality is preserved it is an atom of nickel which is replaced by an atom of lithium. The replacing lithium atom occupies a normal nickel ion site in the lattice, but it becomes only singly charged. Each atom of lithium forming Li^+ contributes only one electron to the crystal, instead of the normal two. Therefore each lithium ion must be accompanied by one electron hole, which can be regarded as Ni^{3+}. In the immediate neighbourhood of the Li^+ ion there is a local excess of negative charge, and the electron hole can circulate round this charged region. It can also escape into the lattice and produce p-type semiconduction. In nickel oxide (which when pure and stoicheiometric is an insulator) the incorporation of lithium ions can increase the conductivity one hundred million times. The defect centre can be described as a foreign cation of lower relative charge on a normal cation site associated with an electron hole on a neighbouring normal cation. The nickel oxide containing lithium ions can be prepared by heating lithium oxide and nickel oxide together in the presence of oxygen. The reaction can be represented by the equation:

$$\tfrac{1}{2}\delta Li_2 O + (1 - \delta)NiO + \tfrac{1}{4}\delta O_2 = (Li_\delta Ni^{2+}_{1-2\delta} Ni^{3+}_\delta) O$$

The oxygen must be allowed to participate in the reaction if it is desired to avoid the production of oxygen ion vacancies, which would lead to n-type semiconduction.

The type of argument given above can be generalised. Incorporation of foreign metal ions of higher valency in the lattice composed of M^{2+} and X^{2-} ions will produce n-type semiconductivity. Incorporation of foreign metal ions of lower valency will produce p-type semiconductivity. Incorporation of foreign electronegative ions of higher valency will produce p-type semiconductivity, and those of lower valency will produce n-type semiconductivity. Semiconductors of the type arising in this general way are called 'controlled-valency semiconductors'. This general type of argument, based only on considerations of charges, neglects all the other factors which may be important in a real chemical system. Although controlled-valency semiconductors can be made on the lines indicated above, it does not necessarily follow that every crystal containing incorporated foreign ions of different valency must be a semiconductor. There may be special factors operating in any particular specific chemical system.

Catalysis by semiconducting ionic crystals and electron transfer at the catalytic surface

Many experimental observations which have been made during the course of researches on heterogeneous catalysis can be interpreted to some extent with the help of the hypothesis that electron transfer at the catalytic surface determines the rate of numerous catalysed reactions. A process determining the rate of reaction may be either the transfer of electrons from the adsorbed layer into the catalyst, or the transfer of electrons from the catalyst to the adsorbed layer. There are two distinct possibilities depending on the direction of electron transfer, either into or out of the catalyst. When the catalytic process requires entry of electrons into the catalyst and this is a semiconductor, a p-type semiconductor best satisfies the electronic requirements for catalysis, since there are electron holes present into which electrons from the adsorbed layer may be transferred. Conversely, for transfer of electrons to the adsorbed layer an n-type semiconductor is most suitable as it contains a number of fairly free and mobile electrons which are readily transferred.

The decomposition of nitrous oxide (dinitrogen oxide) on a number of semiconducting oxides has been extensively studied. For this reaction there is a clear relation between catalytic activity and semiconducting properties. The p-type semiconductors are much better catalysts than the n-type semiconductors, and insulators occupy an intermediate position. Catalysis on insulators is not particularly clarified by the views advanced above, and this serves to emphasise that these views are only an attempt to outline a complex field with the help of simple ideas. Factors other than electron transfer are important in heterogeneous catalysis.

For the reaction between carbon monoxide and oxygen on a variety of metal oxide catalysts a pattern of catalytic activity may be discerned, with the p-type oxides the best catalysts, the n-type oxides being less efficient, and the insulators being even less efficient. Processes of electron transfer at the catalytic surface are probably involved in this oxidation.

Adsorbed particles can donate electrons to a lattice just as donor impurities or donor centres can. When this donation occurs on an n-type semiconductor the conductivity increases. On a p-type semiconductor this donation of electrons causes a decrease of conductivity, since some of the donated electrons will annihilate the electron holes originally present. Similarly, adsorbed particles can accept electrons from a lattice, so that the conductivity of an n-type semiconductor is decreased as adsorption occurs, whereas this increases the conductivity of a p-type semiconductor. One may conceive of a reaction between two reactants, one of which behaves as a donor and the other as an acceptor on the same catalyst. The reactant particles are then drawn together on the surface by electrostatic attraction.

Ideas of the general type outlined in this chapter provide a possible explanation for some of the phenomena shown by catalytic promoters. Foreign ions may have a great influence on the electronic properties of a solid, and it follows from the concepts outlined in this chapter that a foreign ion can affect the electronic properties of the whole lattice. The effect is not necessarily limited to the immediate region of the foreign ion. Another way in which a promoter may operate, in some cases, is by altering the rate of recombination of electrons and electron holes. A practical catalyst, when a semiconductor, may contain defects of several kinds so that both mobile electrons and electron holes contribute to semiconduction. Both electrons and electron holes may be involved in the catalytic process. Foreign atoms, or the surface of a promoter, may provide places where the mobile electrons and holes can be trapped and caused, in effect, mutually to annihilate each other, with a consequent great change in the electronic properties of the solid.

Catalysis by metals and electron transfer at the catalytic surface

Many attempts have been made to discuss catalysis by metals in terms of ideas derived from various electron theories of metals. One of the simplest of these theories is the free-electron theory. A metal, according to this view, is looked upon as a potential hollow containing electrons. The idea of a potential hollow is most easily seen by considering only one dimension of space. Consider an electron which can move only in one dimension. Let this be the x-axis. Suppose that the potential energy of the electron is everywhere infinitely large except for a region between the limits $x = 0$ and $x = a$, where the potential energy is zero. This means that there is an infinitely deep potential hollow of width a. This is represented in Fig. 9. Since the electron must have infinite potential energy to be anywhere except in the region

Fig. 9. A one-dimensional potential box

$0 < x < a$, an electron of only finite energy must be contained within the one-dimensional box. The de Broglie matter waves associated with the electron must then fit exactly into the potential box. According to L. de Broglie a single particle of mass m and velocity v situated in a region free of force has associated with it a wavelength λ given by

$$\lambda = h/mv \tag{4.1}$$

where h is the Planck constant. If the electron is confined to the potential box the associated matter waves must fit exactly into the potential box, and therefore

$$\tfrac{1}{2}n\lambda = a \tag{4.2}$$

where n must be an integer. An integral number of half-waves will just fit into the box with the amplitude of the wave always being zero at $x = 0$ and $x = a$. The energy E of an electron inside the box is entirely kinetic, since the potential energy is zero. Therefore $E = \tfrac{1}{2}mv^2$ and from equations (4.1) and (4.2)

$$E = n^2h^2/8ma^2. \tag{4.3}$$

Since n is an integer, 1,2,3 . . ., the possible energies of the electron are quantised. Equation (4.3) gives the possible energy levels available to a single electron. A basic idea, used in the free-electron theory of metals, is that two electrons of opposite spin may be assigned to each of these energy levels. The assignment is then in accordance with the Pauli exclusion principle. Electrons are fed (in a conceptual process) into the box filling up the lower energy levels first, so that the total energy (the sum of all the individual energies according to this simple model) is always as low as possible. As the number of electrons inside the box is increased, the energy of the most energetic of the electrons will increase. For example, with six electrons the most energetic electrons have $n = 3$, and with eight electrons they have $n = 4$.

   For a solid metal a three-dimensional potential hollow is similarly considered, and again the energy of the most energetic electrons increases as the number of electrons inside the three-dimensional box is increased. The allowed energies for a single electron of mass m in a three-dimensional rectangular box with sides of length a, b and c are

$$E = \frac{h^2}{8m} \left(\frac{n_1{}^2}{a^2} + \frac{n_2{}^2}{b^2} + \frac{n_3{}^2}{c^2} \right) \tag{4.4}$$

where n_1, n_2 and n_3 can only have integral values, 1, 2, 3 . . ., so again the possible energies of the electron are quantised. The numbers n_1, n_2 and n_3 are called 'quantum numbers', and n in equation (4.3) is also a quantum number. In a three-dimensional box the energy of any electron state is specified by the three quantum numbers n_1, n_2 and n_3. These numbers are put into equation (4.4) to give the energy of the electron state. When these numbers are large there are very many states with different values of these quantum numbers but with the same value, or very nearly the same value, for the energy. This is most easily seen by considering a square box with $a = b = c$. Then, for example, the three states

$$n_1 = 3, \quad n_2 = 4, \quad n_3 = 5$$
$$n_1 = 5, \quad n_2 = 3, \quad n_3 = 4$$
$$n_1 = 4, \quad n_2 = 5, \quad n_3 = 3$$

have the same energy.

Equation (4.4) can be deduced by solution of the Schrödinger equation for a single particle of mass m confined in a three-dimensional potential hollow. Equation (4.3) can also be deduced from the Schrödinger equation.

When the free-electron theory is applied to a solid metal, it is again assumed that the metal may be treated merely as a three-dimensional potential box and that two electrons of opposite spin may be assigned to each energy level specified by quantum numbers n_1, n_2 and n_3. As the number of electrons inside the box increases, the energy of the most energetic electrons increases. It is supposed that at 0 K the total energy, which is the sum of all the separate energies of each individual electron, is as small as possible. Therefore the electrons are assigned to the possible energy states so that the levels of lowest energy are always filled first, each with two electrons. The energy of the most energetic electrons in the metal at 0 K is called the Fermi energy (after E. Fermi who originated fundamental ideas in this field). At any temperature above 0 K some of the electrons move, as a result of thermal agitation, into higher energy states, so at temperatures above the absolute zero it is not always the lower energy states which are filled first with electrons. A temperature of 0 K is therefore considered when the Fermi energy is defined.

The free-electron type of theory can only usefully be applied to the valence electrons in a metal — for example, in solid sodium it will only be useful in considering properties arising from one free electron per sodium atom (the 3s electron in the free atom). The other electrons are too tightly bound to the atom to be dealt with usefully by the free-electron theory.

The average number of valence electrons per atom is increased when an alloy is formed by the addition of a metal of higher valency to one of lower valency. For example, silver can be alloyed with cadmium, indium, tin, thallium, mercury, antimony, lead and bismuth. As the element of higher valency is added to the silver, the number of valence electrons per atom is increased. Therefore, on the basis of the arguments above, the Fermi energy also increases as the higher valency element is added. The free-electron concentration (that is the number of free electrons per unit volume) increases. However, the structure of these silver alloys does not change provided that the proportion of higher valency element in the alloy is not too great. Phase rule and X-ray studies on these alloys show that the initial phase (the α-phase, as it is called) formed as the higher valency element is added has the same cubic close-packed structure as silver itself; substitutional solid solution occurs. Indeed, this is what generally happens when copper, silver and gold are alloyed with metallic elements of the B-groups of the second to fifth column of the periodic table.

41

G. M. Schwab and his co-workers in a series of papers (1946-50) found a relation between the number of free electrons per atom in such alloys and the activation energy for the heterogeneous catalytic decomposition of formic acid on these alloys. For example, on pure gold the activation energy was about 11 kcal mole^{-1} (46 kJ mol^{-1}) and this increased to about 27 kcal mole^{-1} (113 kJ mol^{-1}) on alloying 20% of cadmium with the gold. Similarly, the activation energy for the decomposition of formic acid on silver was increased by alloying the silver with the higher-valency elements mentioned above. Observations of this kind reveal some general effect operating independently of the specific chemical nature of the system, and such observations are particularly significant in elucidating the general factors operative in catalysis by metals. The idea of electron transfer from the reacting molecule (or particle) into the metal correlates these observations. Electron transfer into the metal should occur more readily as the Fermi energy decreases. The electrons entering the metal can only enter an electron state not already occupied by electrons. Therefore the first electrons entering the metal must, when in the metal, have an energy infinitesimally exceeding the Fermi energy, and additional added electrons must have still more energy, as soon as all the states with the energy infinitesimally exceeding the Fermi energy are filled with electrons. This argument follows from what has already been said about the filling of the energy levels at 0 K, and it is evidently valid for this temperature. However, at ordinary temperatures, such as those used in catalysis, the occupancy of electron states by electrons does not differ much from that at 0 K, and the argument is still valid. That the occupancy of electron states by electrons in a metal does not change much with temperature gives an explanation of why the specific heat of metals agrees fairly well with the law of Dulong and Petit. The electrons contribute only slightly to the specific heat. This follows immediately from the only slight change of occupancy with temperature.

According to Schwab and his colleagues the activation energy for formic acid decomposition on alloy catalysts increases with the Fermi energy. The hypothesis that the catalytic activation consists in an entrance of electrons from an adsorbed particle into the metal or metal alloy explains this finding. Electron transfer into the metal should occur less readily as the Fermi energy increases. Hence the activation energy of the reaction increases as the Fermi energy increases.

Generally a high Fermi energy is beneficial when the catalytic process involves a transfer of an electron from within a metal to an adsorbed particle, and a low Fermi energy is beneficial when the catalytic process involves the transfer of an electron from an adsorbed layer to the interior of the underlying metal. Many results can be correlated with this simple view. However, we must note that catalytic phenomena are very diverse and complex, and the idea of electron transfer, although useful in a general sense in correlating many observations, cannot, as yet, provide the basis of a comprehensive theory of catalysis. Such a theory is still elusive. E. K. Rideal, a very distinguished worker in catalysis, summarised views he had developed during half a century of research in a book he published in 1968, just fifty years after the early joint book on catalysis he wrote with H. S. Taylor. He noted that the classical example of a heterogeneous catalytic reaction is provided by the hydrogenation of ethylene, and that this still provides an area for vigorous debate. Rideal remarked in his book: 'In the field of catalysis the crucial experiment is often a will o' the wisp and eludes our grasp.' The classical example of a catalytic reaction involving only one reactant is, perhaps, the formic acid decomposition, which is again an area of vigorous debate and renewed experiment, and it may well be that the simple correlations with electron transfer processes deduced for this reaction will need rethinking in the light of new findings.

Holes in the d-band

In metals such as iron, cobalt, nickel and copper, it is possible to regard the outermost electrons in the crystalline metal as being in two bands derived from the 3d and 4s states of the free atom. As the atoms come close together to form a crystal (in a conceptual process) these states broaden into bands. A diagrammatic representation of this broadening is shown in Fig. 10. For nickel, the broadening of the d-band is less than that of the s-band, as shown in this figure. At the internuclear separation in the crystal the s-band completely overlaps the d-band. Electrons are assigned to these bands to keep the total energy a minimum, just as in the free-electron theory. The first electrons go into the s-band, and when a high enough

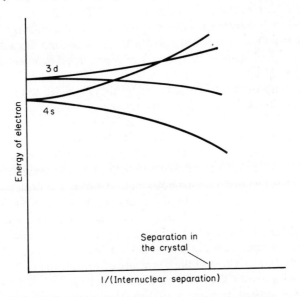

Fig. 10. The d-band and the s-band in a crystalline metal derived from 3d and 4s states of the free atom

energy is reached electrons go into both the s-band and the d-band. The theoretical treatment of ferromagnetism based on this type of theoretical approach relates ferromagnetism to unpaired electrons in the d-band, and from the magnetic properties (and particularly from the saturation intensity of magnetisation) the number of unoccupied electron states in the d-band can be found. One unoccupied electron state accompanies one unpaired d-band electron. The number per atom of unoccupied electron states in the d-band of nickel is 0·6. The number of electrons per atom in the d-band and s-band of nickel can easily be worked out knowing this number of unoccupied electron states. In nickel the $3s^2$ and $3p^6$ sub-groups are full, and we have 18 electrons per atom to divide amongst the 3s, 3p, 3d and 4s states. Since the d-band can hold just 10 electrons per atom, and 0·6 of these are missing, the number present is $(10 - 0·6)$ and the number of 4s electrons is

$$18 - 8 - (10 - 0·6) = 0·6.$$

Here the numbers of 3s and 3p electrons (8) and of d-electrons $(10 - 0·6)$ are subtracted from 18.

It is usual to refer to the unoccupied electron states in the d-band as d-band holes or d-band vacancies, and to refer to these colloquially as 'holes'. The fractional numbers of d-band holes may imply that a particular nickel ion at a given moment has associated with it either a complete sub-group of 10 electrons, or only eight.

In copper, which is not ferromagnetic or paramagnetic, the d-band is fully occupied and there is one electron per atom in the s-band. From the magnetic properties of copper-nickel alloys containing increasing proportions of copper it can be deduced that in these alloys the s-electron associated with a copper atom fills a d-band hole in nickel. In an alloy containing 60 atomic % of copper the d-band holes are just filled.

Studies of a number of reactions catalysed by alloys have led to the view that unpaired electrons in the d-band do, in some instances, play an important part in the catalytic process. When this process involves transfer of an electron into the metal from an adsorbed layer, the electron can enter a d-band hole and the presence of these holes is beneficial for the catalysis. When in the catalytic process a covalent bond is formed between an adsorbed particle and the underlying metal, the unpaired d-electrons associated with the d-band holes may be involved in the bonding and the presence of these holes is again beneficial for catalysis. However, for formation of a negative ion at the catalytic surface d-band holes do not seem to be beneficial. Metals without d-band holes may be the better catalysts in such a case.

Early ideas on the significance of d-band holes in heterogeneous catalysis were put forward by D. D. Eley (then of the University of Bristol) and by D. A. Dowden of Imperial Chemical Industries, and were discussed at a meeting of the Chemical Society in London in 1949. Dowden noted several examples of reactions for which the filling of d-band holes in a metallic catalyst by alloy formation reduced the catalytic activity considerably. Eley with A. Couper found that the activation energy for the reaction on a palladium-gold alloy catalyst of conversion of parahydrogen to orthohydrogen increased abruptly when the holes in the d-band of palladium were filled by electrons from gold alloyed with the palladium. Nearly pure parahydrogen (in which the nuclear spins of the two protons in the molecule are opposed) can be made by reaction of ordinary hydrogen (a mixture of orthohydrogen and parahydrogen) on a charcoal catalyst at very low temperature. The conversion of the parahydrogen to orthohydrogen on a variety of catalysts at higher temperatures has been extensively studied.

Eley and Couper also noted that the d-band of palladium can be filled with electrons from dissolved hydrogen atoms, and again this filling caused an increase in the activation energy for the parahydrogen conversion reaction. Eley remarked that the occurrence of a similar effect with hydrogen dissolved or embedded in platinum might explain why M. Faraday in his early work on the spontaneous combination of hydrogen and oxygen on the platinum electrodes of an electrolysis cell normally observed much greater catalytic activity with the positive electrode of the cell. The negative electrode would have been exposed to hydrogen atoms, and electrons from these atoms would have filled d-band holes in the metal.

5 The possible importance of active catalytic centres at the catalytic surface

The concept of active centres

The idea that on the surface of a solid catalyst there may be certain sites where the catalytic activity towards a given reaction is exceptionally high compared with that on most of the surface has been discussed for many years. Such sites are called active centres. H. S. Taylor in 1925 advanced the hypothesis of active centres in a definite form, and considered that groups of atoms fixed in metastable positions and associated with high energy could form such centres, which might have exceptional catalytic and adsorptive power.

Two generally observed phenomena are explicable with the help of the hypothesis of active centres. First, the heat evolved during the initial stages of the formation of an adsorbed layer on a catalyst surface is usually greater than that evolved during the later stages of the adsorption. In fact, the heat evolved, per mole adsorbed, for each infinitesimal increment of gas adsorbed, usually decreases as the fractional coverage of the surface increases. In experiments in which the heat evolved is measured by a calorimetric method, a small increment of gas can be admitted – so small that the surface coverage does not increase significantly – and the heat evolved is measured and then expressed as a heat evolved for one mole adsorbed on such a large quantity of adsorbent that the surface coverage remains effectively constant during the adsorption. A relation between heat of adsorption per mole (for a given surface coverage) and surface coverage can then be found from experimental measurements. The hypothesis that the surface of the solid contains active centres where the heat of adsorption exceeds that on other parts of the surface explains a fall of heat of adsorption with surface coverage. Such a fall will arise if the adsorbed particles can move over the surface so that they are preferentially adsorbed at those sites where the heat of adsorption is greatest. This will happen because the system as a whole will tend to attain a state of minimum energy (except for a disturbing effect of thermal agitation). To produce the observed effect of a decrease of heat of adsorption with increase of coverage, the adsorbed particles must be able to move over the surface in the time needed to make the calorimetric measurements. If, conversely, the adsorbed particles remain fixed at the first point where they strike the surface, and they strike it at random, the measured heat of adsorption (expressed for one mole, or any constant amount) for each infinitesimal increment adsorbed will not vary with surface coverage although active centres may be present.

The second general phenomenon which is explicable by the hypothesis of active centres is that catalysts are often poisoned by much less poison than is needed to form a monolayer of poison. According to the Langmuir view a poison occupies a surface to a certain fractional extent, and only the part of the surface not covered by poison is available for catalysis. Let the fraction of surface covered by poison be θ_p. Then only the fraction $1 - \theta_p$ of surface is available for catalysis. In a typical poisoning experiment, R. N. Pease observed that mercury appreciably reduced the activity of a copper catalyst for ethylene hydrogenation although considerable adsorption could still occur on the catalyst; the hypothesis that the mercury was adsorbed on a small number of active centres which in consequence were poisoned explained this observation if it was assumed that the adsorption occurred on the remainder of the surface which had little catalytic activity. In another experiment, a copper catalyst which could strongly adsorb 5 cm^3 of carbon monoxide suffered a 90% loss of catalytic activity after adsorbing only 0·05 cm^3 of carbon monoxide. Hence the catalyst seemed to owe 90% of its activity to less than 1% of its strongly adsorbing centres. The hypothesis that the carbon monoxide is preferentially adsorbed at active centres also responsible for the catalysis explains the observed effect. Many similar examples of poisoning have been found.

Alternative theories have been proposed to explain, on the basis of a surface consisting of a uniform set of sites, both the fall of the heat of adsorption with surface coverage and the poisoning phenomenon. Thus, the concept of electron transfer in adsorption implies that charged adsorbed particles are formed, and if these all have the same sign of charge they will repel each other electrostatically. This will evidently cause the heat of adsorption to fall as surface coverage increases. Many of the phenomena of catalytic poisoning can be explained, with a model of a uniform set of sites, by considering the geometrical requirements for adsorption of a molecule on a number of neighbouring atoms, all of which must be free and not covered by poison. Elimination of the adsorptive capacity for a reactant of a single atom by adsorption on it of poison then causes a number of the neighbouring atoms to become ineffective in catalysis. A statistical treatment of this model has been developed, and it explains many of the observed examples of poisoning. We note, therefore, that the arguments for the existence of active centres based on heats of adsorption and on poisoning phenomena are not conclusive.

Terraces, steps, kinks, adatoms, and vacancies at crystal surfaces

Various researches carried out by several investigators round about 1930 on the mechanism of crystal growth developed the view that incomplete ledges or steps on the crystal surface may play an important part in the crystal growth process. Fig. 11 shows a surface with terraces and steps or ledges. The steps are not straight, but con-

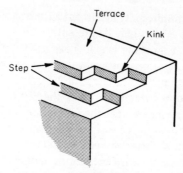

Fig. 11. A surface with steps and terraces, and kinks in the steps

tain kinks, of which three are shown. The steps may be only one atom high (monatomic steps) and a kink in a step may have a size corresponding to only one atom. An extra atom adding on to the crystal is most tightly bound when it adds on at a kink, since it can then interact with the greatest number of neighbours. One mechanism of crystal growth involves the continual adding of atoms or molecules at a kink so that the kink moves along the step until it reaches the edge of the crystal. Another atom must then add at the edge of the complete step, forming, in effect, two kinks which can move to opposite ends of the crystal as crystal growth occurs.

Fig. 12 shows a surface with an adatom and a surface vacancy. For ease of drawing, the atom has been represented as a cube. On the surfaces of a number of metals the departures from ideal structure shown in Figs. 11 and 12 can be seen directly by field ion microscopy.

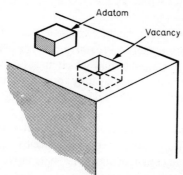

Fig. 12. A surface with an adatom and a surface vacancy

Intersection of edge and screw dislocations with surfaces

A dislocation in a crystal is a structural defect involving large numbers of atoms, in contrast to the point defects already discussed. An edge dislocation can be visualised

48

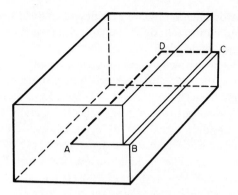

Fig. 13. An edge dislocation is introduced into a perfect crystal by producing unit slip above the plane ABCD. AD is the dislocation line

in the following way. Consider a block of crystal, built from atoms, in which all the atoms are properly arranged, and suppose that part of the crystal above a horizontal plane ABCD (Fig. 13) is pushed in so that a step on the surface is produced, one atom high, running from B to C. Then unit slip has occurred over the area ABCD. It is called unit slip because the displacement is one atom (or more generally one unit of the structural units building the crystal). In the top half of the crystal above the plane ABCD there is one extra layer of atoms. Suppose that this extra layer ends at the line AD so that when this extra layer intersects a surface, for example at A, the arrangement of atoms at the surface is as shown in Fig. 14. In this figure the lines represent rows of atoms in the surface near the point A, and XA represents the extra layer of atoms. There is around A a region of considerable atomic misfit. The misfit introduced by the conceptual process shown in Fig. 13 is concentrated over a small region. Dislocations in metals may be studied directly by field ion microscopy, and

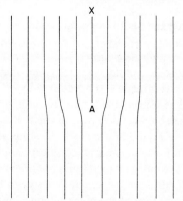

Fig. 14. An edge dislocation which is the edge of the incomplete atomic plane XA

49

less directly by electron microscopy. We have described the dislocation concept for a crystal composed of atoms, but, quite similarly, dislocations may exist in a crystal built from ions or from molecules.

The type of dislocation shown in Fig. 14 is called an edge dislocation and has the feature that the conceptual unit slip which produces it is in a direction normal to that of the dislocation line. The extra layer of atoms ending along the line AD in Fig. 13 would end in a straight edge if it was actually generated by the conceptual process we have discussed. In actual fact, dislocations arise during crystal growth or as a result of mechanical working of a crystal, and there are believed to be monatomic steps in the dislocation edge, like the kinks on a surface ledge. These kinks in a dislocation edge are usually called jogs. At a jog, atoms from the crystal can add to the dislocation edge; if these added atoms leave vacancies behind, we can say that vacancies can leave a dislocation edge at a jog. Similarly a vacancy can be absorbed at a jog which then moves along the edge through a distance corresponding to one atom.

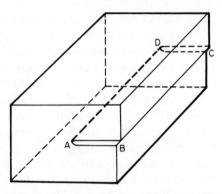

Fig. 15. Production of a screw dislocation by unit slip over ABCD. The screw dislocation is along AD

Atoms can also leave a jog and neutralise vacancies in the bulk of the crystal. The atoms diffuse to the vacancies. Briefly, an edge dislocation jog can act as a sink and as a source for both atoms and vacancies.

Visual experience of the idea of an edge dislocation may be gained by the use of a simple bubble raft. This technique was introduced by W. L. Bragg and J. F. Nye in 1947 and subsequently has appeared in simplified form in a number of books.

A screw dislocation can be visualised with the help of a similar conceptual operation applied to a block of crystal in which all the atoms are properly arranged. However, the unit slip that produces a screw dislocation is in a direction parallel to the direction of the dislocation, as shown in Fig. 15. Here unit slip has occurred over the area ABCD, so along AB there is a monatomic step or ledge on the surface. The misfit of atoms on the front face of the block of crystal is concentrated round the point A, where the layers of atoms are arranged like a spiral staircase, that is like the thread which spirals round a screw. Along the ine AD in the crystal the atoms form a spiral ramp. This is a screw dislocation. The screw and edge dislocations represent extreme types, and intermediate types and curved dislocations are possible. A dislocation is an extended defect, in contrast to a point defect.

Active centres in catalysis

Many experiments with metallic catalysts have shown that catalytic activity can be increased for various reactions by increasing the number of dislocations or of vacancies in the catalyst. For example, I. Uhara and his colleagues, working in Japan, introduced vacancies and dislocations into nickel by mechanically working the metal at ordinary temperatures (cold-working). At higher temperatures (200 to 300°C) the vacancies were annealed out by various processes, for example by diffusion of vacancies to the surface of the crystal or by absorption of vacancies on jogs. At still higher temperatures (400 to 700°C) the dislocations could be annealed out — for example, an incomplete edge can grow by vacancy emission until it reaches a surface. The dislocation associated with the incomplete edge then vanishes. Again, we note that the spiral arrangement of atoms produced by the shearing process shown in Fig. 15 can be in the opposite sense to that shown in this figure. In this figure the part of the crystal above the plane ABCD has been sheared backwards. However, it could just as well have been sheared in a forward direction, putting the ledge facing upwards (AB in Fig. 15) at the back of the crystal block. A spiral arrangement of atoms would again be produced along the line AD. The two arrangements of atoms along AD produced by shearing backwards or forwards above the plane ABCD are related like the threads on two screws, one of which must be turned clockwise to screw it up, and the other of which must be turned anti-clockwise. With screws, one refers to right-handed and left-handed threads. A screw dislocation with atoms spiralling in one sense can be annihilated by meeting a screw dislocation with atoms spiralling in the opposite sense. A dislocation line can move slowly at high temperatures as the atoms in a solid become increasingly mobile as the temperature increases. So screw dislocations can slowly be annealed out.

Uhara and his colleagues showed that the disappearance of vacancies and of dislocations as a result of annealing cold-worked nickel was accompanied by a loss of catalytic activity for several reactions, including reactions of gases on the nickel surface and reactions in solution catalysed at a nickel-solution interface. Many similar experiments have been carried out by other groups of workers.

With semiconducting catalysts, the presence of point defects and of dislocations can cause important alterations in the electronic structure of the solid in the region of the defect. Another effect arises because defects may trap electrons and electron holes. In addition, a catalyst may contain both electrons and electron holes, and the presence of defects of various sorts can assist the process of recombination of electrons and electron holes, which results in mutual annihilation.

Many experiments have been carried out with catalysts which have been subjected to ionising radiation. The term 'ionising radiation' is used with reference to high-energy particles and to high-energy electromagnetic radiation. An atom in a solid can be knocked out of its normal position by a direct collision with a heavy particle. An atom can also be knocked out of its normal position as a result of the recoil imparted to it by the ejected electron when the atom is ionised by an X-ray or γ-ray quantum. Thus ionising radiation produces point defects, notably vacancies and interstitial atoms. However, in an irradiated metal, subsequent processes, such as agglomeration of either vacancy or interstitial point defects, may lead to the formation of dislocations.

Many experiments have shown that irradiation of solid catalysts alters their catalytic activity. Usually the changes of activity are not very large. The exact nature of the active sites involved is still not known. Although field ion microscopy makes it possible to see various kinds of defects at metal surfaces, it has not yet been possible to see chemical reactions actually occurring on a surface and to see how a molecule behaves when it approaches a given surface site. Techniques such as that of the atom probe field ion microscope may in time throw light on this problem.

Experiments

The following suggestions are for experiments which may be carried out as projects. Therefore no detailed instructions are given. The quantities to be used must be found out by the experimenter, as in a research problem, and the experimenter must also devise his own method and, where applicable, his own apparatus.

Adsorption from solution

Many projects can be devised on this topic. A solution of a known concentration is shaken with a finely divided solid which absorbs solute from solution. The decrease of concentration due to this adsorption is found, and the results may be plotted in terms of the Langmuir isotherm.

Charcoal, obtainable from laboratory suppliers in various forms, may be used to adsorb acetic acid (ethanoic acid) and iodine. Charcoal may also be prepared in the laboratory by pyrolysis of cane sugar and of sawdust, and it may be activated by heating in steam. The investigation of the adsorptive properties of charcoal prepared in different ways and activated with steam under different conditions would provide a project.

The adsorption of cobaltous ions (cobalt(II) ions) by finely divided glass provides another problem for project work. Questions which may be investigated are: the effect of pH; whether the adsorption is accompanied by ion exchange, so that (with a soda glass) sodium ions in the glass are replaced by cobaltous ions (cobalt(II) ions); and whether the adsorption gives an electrical charge to a glass particle which then moves in an electrical field. This movement could be investigated microscopically, using light scattered from the particle.

The adsorption of citric acid (2-hydroxypropane-1,2,3-tricarboxylic acid) by freshly precipitated aluminium hydroxide may be studied to elucidate the effects of conditions of precipitation, of ageing of the precipitate, and of collecting and drying the precipitate before it is used as an adsorbent.

Similarly the adsorption of chromic salts (chromium(III) salts) by freshly precipitated barium hydroxide may be studied.

Catalytic decomposition of hydrogen peroxide in aqueous solution

Many experiments may be devised to illustrate, with this decomposition, both homogeneous and heterogeneous catalysis. The decomposition catalysed by bromide ions shows an effect of neutral salts. The decomposition may be catalysed by cupric (copper(II)) and other metal ions, and by mixtures of ions, e.g. cupric (copper(II)) and ferric (iron(III)) ions. The decomposition is also catalysed by colloidal metal sols. Such sols could be prepared by the arc method or in other ways.

Other solution reactions

The effect of neutral salts in aqueous solution on the rate of oxidation of iodide ions by persulphate ions (peroxodisulphate ions) provides a possible project. The liberated iodine is easily estimated. The oxidation of bromide and iodide ions by bromate ions may also be investigated in the presence of various salts.

The decomposition of ammonium nitrite in aqueous solution is reported to occur at a reasonable rate at about $70°C$. A suitable solution can be prepared by mixing cold saturated solutions of sodium nitrite and ammonium chloride. A search for homogeneous and heterogeneous catalysts for this reaction would provide a project. The reaction rate can be followed by measuring the evolved nitrogen.

Gas reactions catalysed by metals

The oxidation of methanol to formaldehyde catalysed by a platinum wire can be demonstrated as originally described by H. Davy. A few drops of methanol are placed in a wide-mouthed flask warmed on a water bath. A heated platinum wire introduced into the mixture of methanol vapour and air in the flask will, if conditions are right, become incandescent and remain so for a long time. Experiments on catalyst poisoning and catalyst activation can be devised with this reaction.

The dehydrogenation of methanol occurs readily on a heated copper catalyst. The methanol vapour may be passed in a stream of nitrogen over copper heated in a glass tube, and formaldehyde estimated in the issuing gas. It would be possible to investigate the use of carbon dioxide (from solid carbon dioxide) as an alternative carrier gas.

A catalysed chemiluminescent reaction

A project could be devised involving the preparation of iron phthalocyanine and its use to catalyse the oxidation of luminol (NN'-3-aminophthaloylhydrazine) in solution. Luminol is commercially available. The light emission is visible in daylight with the right conditions.

Catalysis in heterogeneous systems in an aqueous medium

The reaction

$$C_2H_5I + AgNO_3 + H_2O \rightarrow C_2H_5OH + AgI + HNO_3$$

is catalysed by charcoal. It is also catalysed by silver iodide, formed by the reaction. In devising a project to study this reaction, care must be taken not to allow too much volatilisation of ethyl iodide from the solution.

The decomposition of a paste of bleaching powder and water is catalysed when cobalt chloride (cobalt(II) chloride) solution is added. The investigation of this catalysis would provide a project.

The cylindrical field emission microscope

An ambitious project involves the construction and use of a simple cylindrical field emission microscope. Practical details can be obtained from the author.

Bibliography

A clear and brief general elementary introduction to heterogeneous and homogeneous catalysis is given by:

BOND, G. C. *Principles of Catalysis.* Royal Institute of Chemistry Monographs for Teachers, No. 7, 2nd Edition, 1968

A more advanced general introduction to heterogeneous catalysis is given by:
THOMSON, S. J., and WEBB, G. *Heterogeneous Catalysis.* Oliver and Boyd, Edinburgh and London, 1968

A more advanced book giving discussions of research problems in heterogeneous catalysis is:
THOMAS, J. M., and THOMAS, W. J. *Introduction to the Principles of Heterogeneous Catalysis.* Academic Press, London 1967

An introduction to defects in crystals is given by:
REES, A. L. G. *Chemistry of the Defect Solid State.* Methuen, London, 1954

A more advanced treatment of the solid state is given by:
HANNAY, N. B. *Solid-State Chemistry.* Prentice-Hall, Englewood Cliffs, New Jersey, 1967

Many of the topics discussed in this present monograph have been discussed more fully in:
ROBERTSON, A. J. B. *Catalysis of Gas Reactions by Metals.* Logos, London, 1970

The importance of catalysis in industrial practice is illustrated in the following two monographs:
SAMUEL, D. M. *Industrial Chemistry – Inorganic.* Royal Institute of Chemistry Monographs for Teachers, No. 10, 1966
SAMUEL, D. M. *Industrial Chemistry – Organic.* Royal Institute of Chemistry Monographs for Teachers, No. 11, 1966

Index

DATE DUE